Making Scientific Discoveries

Jan G. Michel (ed.)

Making Scientific Discoveries

Interdisciplinary Reflections

BRILL
MENTIS

Printed with the kind support of the DFG as part of the Emmy Noether Research Group »Theology as Science?!«

Funded by

DFG Deutsche
Forschungsgemeinschaft
German Research Foundation

Cover illustration: Jan G. Michel, Radio image of the black hole Pōwehi by Event Horizon Telescope on 10th April 2019 (release), located in Messier 87. / wikimedia.org [https://commons.wikimedia.org/wiki/File:Black_hole_-_Messier_87_crop_max_res.jpg]

Bibliographic information published by the Deutsche Nationalbibliothek

The Deutsche Nationalbibliothek lists this publication in the Deutsche Nationalbibliografie; detailed bibliographic data available online: http://dnb.d-nb.de

www.mentis.de

Cover design: Jan G. Michel
Production: Brill Deutschland GmbH, Paderborn

ISBN 978-3-95743-210-0 (paperback)
ISBN 978-3-95743-704-4 (e-book)

Contents

Preface

Jan G. Michel

As the title slyly suggests, this book gathers articles from researchers across disciplines to reflect on what we do when we make discoveries in science. It grew out of my first tentative attempts several years ago to shed new light on scientific naming processes and the different functions of kind terms. My thinking about this led me to ask what exactly happens when we make scientific discoveries, for example in biology. After discussing this and related questions with colleagues from various disciplines, giving a number of talks, and publishing some articles, it turns out that my original tentative attempts described above marked the starting point of a path that I had not foreseen and that eventually led me to the discovery of a new philosophical field of research: the philosophy of scientific discovery. Something we see particularly clearly these days is that scientific discoveries are not only central to science, but to society as a whole. I believe that this makes it even more important to better understand what constitutes scientific discoveries. That is the main purpose of this book.

I am glad and grateful to have the opportunity to explore this new exciting and inspiring field of research here. Of course, this book would not have been possible without the support of a number of people. First of all, I would like to thank all the contributors to this book and to the conference from which the book emerged: Hanne Andersen, Kim Boström, Jörg Friedrich, Mitch Green, Susanne Hahn, Mario Livio, Joe Moore, Gernot Münster, Theo Michael Schmitt, Niko Strobach, and Christian Tapp. Moreover, I would like to thank Frank Meier-Hamidi of Franz Hitze Haus, Münster, where the conference was held in December 2019, and also Roland Mikosch (camera & production) and Kim Boström (music) with whom I prepared a conference video.[1] Furthermore, I would like to thank all my colleagues who supported me in this project, especially Benny Göcke, Bill Lycan, and Michael Pohl. Research for this project was funded by the German Research Foundation (Deutsche Forschungsgemeinschaft, project number 295845819).

A special thank you goes to Michael Kienecker from mentis, who has accompanied me on a book project for the seventh time now and who has again supported all of my ideas – as always, it has been my great pleasure to work with

---

1 https://www.youtube.com/watch?v=gTh7e3JnNxo

him. Finally, I would like to thank Uta and Hannah; I greatly appreciate the way they have supported me in the middle of a pandemic working on this book. I certainly couldn't have done it without them.

May 2021,
Jan Michel

Making Scientific Discoveries. Editor's Introduction

Jan G. Michel

Scientific progress depends crucially on scientific discoveries. Yet the topic of scientific discoveries has not been central to debate in philosophy. This volume aims to remedy this shortcoming. Based on a broad reading of the term »science«, the volume convenes experts from different disciplines and fields who reflect upon several intertwined questions connected to the topic of making scientific discoveries.

Among these questions are the following: What are the preconditions for making scientific discoveries? What is it that we (have to) do when we make discoveries in science? What are the objects of scientific discoveries, how do we name them, and how do scientific names function? Do discoveries in, say, physics and biology, share an underlying structure, or do they differ from each other in crucial ways? Are other fields such as environmental studies and theology loci of scientific discovery? What is the purpose of making scientific discoveries? Explaining nature or reality? Increasing scientific knowledge? Finding new truths? If so, how can we account for instructive blunders and serendipities in science?

In the light of the above, the following is an encompassing question of this volume: What does it mean to make a discovery in science, and how can scientific discoveries be distinguished from non-scientific discoveries? In what follows, I provide a brief overview of the contributions to this volume.

Jan G. Michel, »Toward a Philosophy of Scientific Discovery«

Jan G. Michel argues that we need a philosophy of scientific discovery. Before turning to the question of what such a philosophy might look like, he addresses two questions: Don't we have a philosophy of scientific discovery yet? And do we need one at all? To answer the first question, he takes a closer look at history and finds that we have not had a systematic philosophy of scientific discovery worthy of the name for over 150 years. To answer the second question, Michel puts forward three arguments that show the importance of a philosophy of scientific discovery. Briefly, he arrives at the following answers: No, we don't yet have a philosophy of scientific discovery, and yes, we definitely need one. To remedy this shortcoming, Michel analyzes the concept of discovery, leading him to the insight that scientific discoveries have an underlying structure

© BRILL MENTIS, 2022 | DOI:10.30965/9783957437074_002

with certain structural features. Some of these features may be important but not indispensable to scientific discovery processes; these include eureka moments, serendipities, joint discoveries, special science funding, and others. In addition, Michel identifies three indispensable structural features which he examines in detail and which he places in a picture with a certain dynamics according to which the process of making scientific discoveries can be seen as a path, leading us from finding and acceptance to knowledge.

Mario Livio, »On the Role of Brilliant Blunders in Science«

Based on his previous work (Livio 2013), the astrophysicist and author Mario Livio turns to an important but often overlooked aspect of scientific processes: scientists sometimes make mistakes. Following Livio, we can distinguish between different kinds of mistakes. While there are inexcusable mistakes that result from scientists being sloppy, careless, or inexperienced, there are also mistakes that result from scientists thinking outside the box. Livio focuses on the latter kind, speaking of »brilliant blunders«. With the help of well-chosen examples from both the history of science and contemporary research, he shows in an impressive way how brilliant blunders can be portals to scientific discoveries and breakthroughs, and that even the greatest scientists are not above error. In line with Michel's picture of scientific discovery processes (this volume, see above), Livio argues that we should not think of scientific discovery and research in general as a linear and exclusively triumphant march toward scientific knowledge, but rather as a zigzag course along which we may sometimes be misguided or led astray. Livio also notes that academic institutions are still quite slow in recognizing that it is almost impossible to further scientific progress without blunders. This, Livio demands, is also why we should allow for originality and creativity in science and why we should make space for thoughtful, if risky, scientific proposals in grant applications and in the evaluation of research projects.

Gernot Münster, »Discovering, Inventing, Contriving – What Are Scientists Actually Doing?«

Maintaining the character of the original public lecture from which it has evolved, Gernot Münster's contribution deals with discoveries in the natural sciences in general, with many of Münster's examples coming from his own field, i.e., from physics. As Münster points out, the term »discovery« can be used to refer to the process of discovery or to the result of it. While philosophy of science discussions about discoveries – to the extent that these discussions exist at all – have mostly been concerned with the process of discovery, Münster is interested in the results of discovery and their characteristics. He wants to find out which kinds of discoveries there are and sketches a draft of a taxonomy of discoveries in the natural sciences. While Münster admits that this might not be much more interesting than stamp collecting, he also takes a look at what actually constitutes a scientific discovery along the way to his taxonomy. In addition, he discusses some related topics in the philosophy of science and eventually even makes an attempt to specify what science is at all: Münster sees science as a collective enterprise which – against a background of accepted background knowledge – aims at rationally defensible, empirically, or theoretically verifiable knowledge, the relevance of which goes beyond the individual case, and to communicate this knowledge to the community.

Niko Strobach, »How Much Realism Does the Word ›Discovery‹ Presuppose?«

Given the observation that the word »discovery« already contains a certain image, namely the lifting or removal of an opaque cover so that what was previously invisible underneath becomes visible, Niko Strobach is concerned with the question if there is a link between the word »discovery« and the opposition between realism and anti-realism in the philosophy of science. Strobach, a realist himself, claims that there is even a strong link. He argues that the word »discovery« already presupposes quite a bit of realism and that it is a word the realist can seriously use, while the serious anti-realist should refrain from using it. Strobach begins his argument with a rough characterization of realism and anti-realism, before he turns to metaphors and provides some information on words in different languages. Strobach then attempts to clarify the logical syntax of »discovery« and »to discover«, distinguishing between an objectual and a propositional use: occurrences of »to discover« followed by a direct object (»America«, »the atom«) should be distinguished from occurrences followed by a that-clause. Following Strobach, realists seem to have some affinity to direct object constructions and anti-realists to that-clauses. Since language is flexible, however, Strobach brings forward some ontological considerations to provide more clarity. He concludes by showing that realists can also use »to discover« in true sentences, even if it is an achievement verb.

Kim J. Boström, »Making Discoveries in Physics«

Elaborating on the notion of scientific discovery as introduced by Michel (2019, 2020, this volume), Kim J. Boström analyzes several discoveries in physics in his contribution. In doing so, he proposes some new terminology to facilitate the distinction and identification of certain types of discoveries:

1.   abstract (or theoretical) versus concrete (or empirical) discoveries,
2.   derived versus non-derived discoveries,
3.   joint discoveries, and
4.   null discoveries.

Boström is optimistic that this is helpful in analyzing real scientific discovery cases. In particular, he emphasizes the importance of accepting certain discoveries as joint discoveries. In doing so, Boström opposes a traditional and more implicit attitude according to which there can only be one discoverer of one and the same thing – even if there are good reasons to believe that there has been more than one discoverer in a given case. Moreover, Boström argues that what he calls the null discovery deserves more attention and appreciation; after all, he says, it has often been shown that the scientific discovery that something is not the case can be just as important as the positive case. Boström thus diagnoses a problem we are faced with in the scientific literature in general, namely that null results, which in many cases really are null discoveries, are inherently difficult to publish, probably because they do not seem sensational enough. In Boström's view, this should change.

Michael Schmitt, »Discoveries in Biology«

Michael Schmitt's contribution starts from the observation that there are different kinds of discoveries in biology. These include the following: causations of empirically observed phenomena, law-like relationships in nature, and species previously unknown to science. To get a better idea of discoveries in biology, Schmitt describes five examples:

1.   the so-called Darwin-finches and their impact on the development of Charles Darwin's theory of evolution,
2.   the viviparous shark of Aristotle and its (re-)discovery by Johannes Müller,
3.   the discovery of ultrasound detection in moths by Friedrich Schaller,
4.   the discovery of the »new insect order« Mantophasmatodea by Klaus-Dieter Klass and co-authors, and
5.   the predicted discovery of the Madagascan hawk moth *Xanthopan morganii praedicta.*

After discussing each of these biological discoveries, Schmitt turns to the general question of what constitutes a scientific discovery. He sketches a picture of scientific discovery that complements Michel's (2019, this volume) in several respects. Among other things, Schmitt emphasizes the role of expectations, that scientists form on the basis of their respective background knowledge, and their interplay with new observations. Moreover, Schmitt claims that a scientific community will accept a finding as a new discovery only if the finding is a genuine novelty and not trivial. Finally, from a broader perspective, Schmitt concludes that science is not done in a sterile environment, but that the social, economic, and political contexts matter as well.

Mitchell S. Green, »Discoveries in Linguistics: Making, Finding, or Somewhere in Between?«

In his contribution to this volume, Mitchell S. Green is concerned with the extent to which scientific discovery is a social process, focusing on discoveries, or putative discoveries, in linguistics as a test case. With reference to a distinction between what is known by an individual and what is known by a group, he argues that the notion of scientific discovery needs to be understood in relation to what is known by a group. He also argues, however, that it is possible for an individual or group to make a scientific discovery without this fact being acknowledged by the scientific community. On this basis, he challenges a view that has been recently brought forward and defended by Michel (2019) that to be a scientific discovery, an intellectual advance must be accepted by relevant experts in the scientific community. This challenge in turn provides reason for doubting that scientific discovery is helpfully analyzed from the perspective of institutional reality as that notion has been propounded in the work of Searle (1995, 2010).

Jörg Phil Friedrich, »The Discovery of Anthropogenic Climate Change«

Jörg Phil Friedrich is concerned with a special case of scientific discovery, namely that of anthropogenic climate change. In his contribution, Friedrich wants both to clarify what constitutes the discovery of anthropogenic climate change and to show how this discovery sheds light on other kinds of scientific discoveries. After identifying three different kinds of scientific discoveries – the discovery of effects in experiments, surprising consequences in theories, and signals in empirical statistics –, Friedrich addresses the question of where and when anthropogenic climate change was discovered. An answer to this question depends, as Friedrich shows, on what we mean by the term »anthropogenic climate change« – do we mean, for instance, the effect of warming air with a higher concentration of carbon dioxide or do we rather mean the consequence of burning coal and oil by humans? Friedrich notes that, from today's perspective, climate change is regarded as an abrupt, radical, and fast change of the whole climate system, including oceans, ice shields, vegetation, and society – it is humanity's greatest challenge. Friedrich concludes that anthropogenic climate change, as we use the term today, has only been discovered over the last decades, mostly with the help of computer simulations, as the statistical signal in the hockey stick curve, and to a certain extent in the weather events of the last few years.

Joseph G. Moore, »Discoveries in Environmental Studies?«

Environmental studies is a new and increasingly popular field of academic study. In his contribution to this volume, Joseph G. Moore wonders what kind of study it is exactly. Environmental studies is often presented as a multidisciplinary field that brings the natural sciences, social sciences and humanities together to study the complex interactions between humans and nature. Given our shared alarm about the state of the global environment, this topic-focused characterization works well institutionally. It draws students, attracts faculty, excites administrators and gives everyone a good enough sense of what goes on in the field. Nevertheless, it is worth asking whether there is something methodologically or otherwise distinctive about environmental studies, something that gives the field the intellectual legitimacy of an established discipline. Is the field anything more than the sum of its disciplinary parts? Does environmental studies offer explanations, understandings or even better, discoveries that are interestingly synthetic? And if so, what would such discoveries look like? Do they flow from environmental studies holistically, or merely from cobbling together the findings of the field's sub-disciplines? Here Moore argues that there are discoveries that fall centrally within environmental studies, and he suggests that these discoveries are guided by an implicit disciplinary lens – »ecoholism« – that can usefully unify the field.

Christian Tapp, »New Findings in Old Religions? On the Possibility of Scientific Discoveries in Theology«

Are there scientific discoveries in theology? Many critics think such a thing is impossible because theology is bound to fixed religious »truths«. In contrast to that, Christian Tapp argues in his contribution to this volume that there actually are scientific discoveries in theology, and he sheds some light on their nature in three major steps. First, Tapp argues, many theological discoveries qualify as scientific as a result of their status as discoveries in other, secular academic fields. Second, systematic theology (which is central to the larger field of theology) admits of hermeneutical and doctrine-related discoveries. Their nature is illuminated in an extensive taxonomy, which suggests counting them as scientific. Third, when the theological subdisciplines contribute to theological discourse, they may establish a new state of research or even a new theological consensus. This new state of research or consensus, in turn, may influence the epistemic enterprises in these theological subdisciplines. This interdependence between particular findings in theological subdisciplines and the general state of theological discourse justifies regarding the contributions in question as genuinely theological discoveries.

Susanne Hahn, »Bullshit in Science? – On Epistemic Norms, Credibility and the Role of Science in Society«

In his book *On Bullshit*, Harry Frankfurt (2005) deals with the phenomenon that people make statements without taking into account the truth or falsehood of what they have said. On the one hand, this happens strategically, for the realization of certain goals, but also, on the other hand, rather carelessly. The latter often happens when people are urged to express themselves without having the necessary knowledge. The question Susanne Hahn is concerned with in her contribution to this volume is whether there is any evidence of the latter type of bullshit in science: Do scientists (occasionally or even often) go beyond what they can correctly say on the basis of their research? What consequences would such a finding have for the special status of the science system in liberal democracies, in which the production of knowledge and its transmission are essential foundations of political participation?

References

Frankfurt, H.G. 2005. *On Bullshit*. Princeton: Princeton University Press.

Livio, M. 2013. *Brilliant Blunders*. New York: Simon & Schuster.

Michel, J.G. 2019. »How Are Species Discovered? Declarative Speech Acts in Biology.« *Grazer Philosophische Studien* 96 (3): 419–441.

Michel, J.G. 2020. »Could Machines Replace Human Scientists? Digitalization and Scientific Discoveries.« In *Artificial Intelligence: Reflections in Philosophy, Theology, and the Social Sciences*, edited by B.P. Göcke & A. Rosenthal-von der Pütten, 361–376. Paderborn: mentis/Brill.

Searle, J.R. 1995. *The Construction of Social Reality*. London: Penguin

Searle, J.R. 2010. *Making the Social World: The Structure of Human Civilization*. Oxford: Oxford University Press.

Toward a Philosophy of Scientific Discovery

Jan G. Michel

»Discovery is what science is all about.«

(Hanson 1967, 352)

2.1 Introduction[1]

The purpose of this essay is to argue that we need a philosophy of scientific discovery and to suggest what such a philosophy might look like. Of course, one may ask whether we don't have a philosophy of scientific discovery at hand already, and one may also call into question whether we really need such a philosophy at all. These concerns are reasonable and we begin by examining each of them in turn. This leads us to the following results: First of all, it leads us to the observation that, from a systematic perspective, the topic of scientific discovery has received little to no attention in philosophy for quite a long time, namely roughly since the middle of the 19th century (§ 2.2). In addition, it leads us to three arguments for the thesis that discoveries play a crucial role for modern science (§ 2.3).

In the light of these findings, we want to gain a better understanding of what happens when we make scientific discoveries. We begin by exploring different cases of discovery (§ 2.4), which brings us to acknowledge that scientific discoveries have an underlying structure with certain structural features, three of which are indispensable or essential to scientific discovery processes. In the remainder of this essay, we take a closer look at these features and their interplay (§ 2.5). Briefly put, we find that the process of making scientific discoveries can be seen as a path, leading us from finding and acceptance to knowledge. This, however, is only the first step toward a philosophy of scientific discovery.

---

1   Preliminary remark: My line of argument in this paper is based on a broad reading of the term »science«, similar to the German term »Wissenschaft«. This reading covers, on the one hand, disciplines we usually regard as belonging to the realms of the natural, social, and formal sciences. On the other hand, it also allows for disciplines which we usually regard as belonging to the realm of the humanities to be scientific. What is more, this broad reading of »science« does not preclude (parts of) other academic fields from being scientific.

2.2   The History of the Philosophy of Scientific Discovery

It is reasonable to ask whether we don't already have a philosophy of scientific discovery – if only because it is an obvious truth that it is characteristic of modern science that, again and again, we make new discoveries in it. Surprisingly, however, scientific discoveries have not yet been a subject of intensive and sustained research, so that we still do not have a full-fledged philosophy of scientific discovery.[2] And the situation is even worse: As I have claimed elsewhere, »the topic of scientific discovery has received almost no attention in philosophy over the course of the last thirty years at least« (Michel 2019, 420). Striking as this may already be, my subsequent research on that question has led me to extend my claim in this essay as follows: The topic of scientific discovery has received hardly any attention in systematic philosophy since the mid-19th century.

This may sound like a bold claim. It is, however, a diagnosis that has been shared by some of the most renowned philosophers and historians of science. For instance, in his »pathbreaking book full of wisdom concerning science« (Hausman 2003, 423), *Science, Truth, and Democracy*, Philip Kitcher writes:

> »Descartes and Bacon continue to shape – and to limit – contemporary discussions of ›proper‹ inquiry. From the seventeenth century to the present, their main questions – What is the right method of discovery? What is the logic of justification? – have dominated reflections about the sciences. In the twentieth century, the first question fell out of favor as an influential group of philosophers argued that there was no general method of scientific discovery. [...] For the past seventy years, however, the central normative question about science has concerned the logic of justification. [...] It is remarkable how much this conception leaves out.« (Kitcher 2001, 109–110)

Kitcher's book was published in 2001, so when he speaks about the »past seventy years«, he refers to the period from around 1930 to 2001. This also fits well with Kitcher's parlance of »an influential group of philosophers«, by which he refers to the adherents of Logical Empiricism and Logical Positivism in the 1920s and 1930s. In a similar vein, Thomas Nickles observed in 1980 already that

> »during the past century philosophers of science, on the whole, have addressed scientific discovery only to exclude it from their domain of professional interest and competence.« (Nickles 1980c, 1)

---

2   This is not to say that there have been no attempts to change this, e.g., cf. Nickles 1980a, Nickles 1980b, and Meheus & Nickles 2009. My claim, however, is that these and other works have not received their due recognition in philosophy of science debates.

So, while Kitcher holds that the topic of discovery fell out of favor in philosophy of science around 1930, Nickles claims that it had been excluded even earlier – somewhere between 1880 and 1900 (depending on what Nickels means by »during the past century«). Beyond that, in asking »Why did traditional philosophy of science ignore discovery?«, Aharon Kantorovich (1993, 97) expresses his agreement with Kitcher's and Nickles' claims and, at the same time, gives voice to the obvious question.

Although my claim that the topic of scientific discovery has received hardly any attention in systematic philosophy since the mid-nineteenth century finds support in the writings of these and other authors, there is still much to be said on that claim in order to clarify and justify it. By providing a brief overview of the deplorable history of the philosophy of scientific discovery over the course of the last more or less 200 years, I want to illustrate how the first half of the nineteenth century saw an original and fruitful systematic philosophical reflection of the concept of scientific discovery inspired by the advancement of science at the time. This presentation, however, will also illustrate how comparatively little has happened afterwards.

My overview is divided into eight parts; it will lead us from John Herschel's forgotten philosophy of discovery (§ 2.2.1) to the work of William Whewell (§ 2.2.2) and his debate with John Stuart Mill (§ 2.2.3), and from there to the influence Herschel's philosophy had on Charles Darwin's work (§ 2.2.4). We will, then, see how the philosophy of scientific discovery has been on the decline after the 1850s (§ 2.2.5), learn that – against the common impression – the so-called context distinction was not original with Reichenbach and Popper (§ 2.2.6), and find that Hanson's logic of discovery is unfortunately underrated (§ 2.2.7). I conclude by taking a look at the philosophy of scientific discovery today (§ 2.2.8) and diagnose: There is currently no sustained and intensive systematic philosophical research on the topic of scientific discovery.

2.2.1   *John Herschel: A Forgotten Philosopher of Discovery*

Let us begin with England's leading and most famous scientist from about 1830 to 1860 (see Cannon 1961, Cannon 1967) – or, more precisely: with England's leading and most famous *natural philosopher*, as this was the customary term at that time. We are talking about Sir John F.W. Herschel (1792–1871). In case you are puzzled or not entirely sure who I am referring to, feel reassured: you are not alone. As Walter Cannon put it, there is a

> »great disparity between John's towering reputation in the mid-19th century and the widespread ignorance of his name after 1900. In my experience, three out of four people who have heard of ›Herschel‹ at all will assume that you have

confused the name of his father, William Herschel, and the fourth is himself not sure about the difference.« (Cannon 1970a, 731; see also his additional remark in Cannon 1970b, 128)

In the light of this, a biographical note seems to be in order:[3] The only child of William (1738–1822) – famously being credited as discoverer of planet Uranus – and nephew of comet discoverer Caroline (1750–1848), John Herschel was born into and raised by a family of renowned astronomers with an inspiring passion for science.[4] Yet, he also faced challenging times as a kid, including being »on the receiving end of bullying« when at Eton College (Cudnik 2013, 27). After being removed from Eton by his mother, he was tutored at home by a private teacher who coached him so well that he was able to enter St. John's College, Cambridge, at the age of 17 already.

As a Cambridge student, John Herschel made friends with the mathematicians Charles Babbage, the father of the computer (Halacy 1970), and George Peacock, later a famous theologian. The three aimed at replacing the Newtonian notation for differentiation by the Leibnizian, to which end they founded the so-called Analytical Society in 1812. In his Cambridge years, Herschel also met the famous polymath William Whewell with whom he shared a lifelong friendship and to whom we will turn below. Herschel's career at Cambridge was a sustained triumph, his aunt recorded that »from Matriculation to Graduation, he gained, without exception, all the first prizes for which he was eligible« (Simenhoff 1963, 122f.). In 1813, Herschel received much admiration for graduating Senior Wrangler from mathematics at Cambridge,[5] making him a natural candidate for the Lucasian Chair of Mathematics, formerly held by Isaac Newton.[6] The same year, the Royal Society published Herschel's first paper and also elected him a Fellow – an exceptional honor, especially at only 21 years of age.

Contrary to what Herschel had originally intended, he decided to assist his renowned father in his astronomical research as of 1816. In doing so, he benefited from his father's unrivaled experience and learned everything about the construction and use of large telescopes, a training that also contributed to

---

3 The most important sources for my presentation include Günther Buttmann's and Eileen Shorland's captivating biographies of John Herschel (Buttmann 1965, Shorland 2016).

4 For a vivid depiction of William Herschel's passion for astronomy beside his being a professional musician, see Tom Standage's book *The Neptune File* (2000). See also Hoskin 2005.

5 Herschel's friend Peacock came out as Second Wrangler, and »[in] spite of the strain of [the] exams, both men sat for the Smith's Prize and again the order was the same« (Shorland 2016, 25).

6 See Buttmann 1965, 23, and Field 1871, 217. Among the more recent Lucasian Professors famously feature Paul Dirac and Stephen Hawking.

his subsequent achievements. In 1820, Herschel was, alongside Babbage, one of the founders of the Royal Astronomical Society. Apart from his astronomical work in these years, however, Herschel made important contributions to chemistry, the physics of light, and also mathematics, for which he earned the Royal Society's highest honor, the Copley medal, in 1821. Herschel served as secretary of the Royal Society, received several prizes, and was knighted in 1831.

After completing his father's catalogue of the northern sky, Herschel and his growing family set sail for the Cape of Good Hope where he spent the years 1834 to 1838 cataloging the southern sky. Among many other things he did there, he recorded the locations of 68,948 stars, discovered over 3,000 double stars and 500 nebulae, gave detailed descriptions of the Magellanic Clouds, and observed Halley's Comet as well as the satellites of Saturn. Apart from his exceptional astronomical achievements at the Cape of Good Hope (which earned him a second Copley medal in 1833), Herschel produced camera lucida drawings of the wild Cape flora which his wife Margaret finished in watercolor, eventually leading to a high-quality coffee-table book, *Flora Herscheliana* (Warner & Rourke 1996). Upon his return to England in 1838, Herschel was celebrated by the scientific world and made a baronet.

Building on a chemical discovery he had already made in 1819 and that had gone largely unnoticed, namely that hyposulphite of soda dissolved otherwise insoluble silver salts, Herschel, independently of Fox Talbot, invented a photographic process on sensitized paper in 1839. In addition, he coined the terms »photography« as well as »positive« and »negative« to refer to photographic images. None less than Charles Darwin was one of Herschel's most prominent admirers. Shortly after visiting Herschel at the Cape in 1836, he wrote in a letter that Herschel »appears to find time for everything« (Barlow 1967, 116). Darwin hit the nail on the head with this.

Now, apart from his outstanding work and discoveries in such diverse fields as mathematics, astronomy, optics, chemistry, meteorology, and even physical geography, Herschel was also a remarkable philosopher – albeit, strangely, a forgotten one.[7] In 1831, Herschel's main philosophical work was published,

---

7   Curiously, this is also the title of Eileen Shorland's fascinating biography of John Herschel: *The Forgotten Philosopher* – a limited print run published by The Herschel Family Archive. I'm very grateful to Will Herschel-Shorland for providing me with a copy of Shorland's book and also for his wonderful support. Will Herschel-Shorland is – like Eileen Shorland, his grandmother – a direct descendant of Sir John and also the editor of Shorland's book (which she had originally written in the late 1960s, see Shorland 2016, Foreword). Yet, as to why John Herschel is a forgotten philosopher today, we can only speculate. According to some, the main reason Herschel is overlooked as a significant philosophical contributor – when in fact

A Preliminary Discourse on the Study of Natural Philosophy,[8] providing an analysis of the roles of hypothesis, theory, and experiment in science – a book that strongly influenced the views of Whewell, Mill, Darwin, and others. Herschel's philosophy of science is often regarded as standing squarely in a Baconian empiricist tradition, and not only the fact that a portrait of Francis Bacon is used on the title page of Herschel's book seems to suggest this reading, but also eulogies as the following:

> »It is to our immortal countryman Bacon that we owe the broad principle and the fertile principle; and the development of the idea, that the whole of natural philosophy consists entirely of a series of inductive generalizations [...]. Previous to the publication of the Novum Organum of Bacon, natural philosophy, in any legitimate and extensive sense of the word, could hardly be said to exist.« (Herschel 1831, 104f., §§ 96 & 97)

According to Baconian inductivism, we arrive at laws and theories by applying certain inductive rules on observations. Herschel, however, saw that many scientific discoveries do not fit this pattern and, by way of looking at example cases,[9] departed from this view. He claimed that there is another way to arrive at laws and theories, namely by creative hypothesizing, a creative process not governed by rules. From Herschel's perspective, both of the following are possible: we can arrive at a theory by means of careful examination of the phenomena or by a sudden flash of inspiration – the actual process of generating a theory being irrelevant to its justification.

> »In the study of nature, we must not, therefore, be scrupulous as to *how* we reach to a knowledge of such general facts: provided only we verify them carefully when once detected, we must be content to seize them wherever they are to be found.« (Herschel 1831, 164, § 170)

In doing so, Herschel drew a distinction between the way things are discovered and the way they are investigated and established – a distinction that, as we will see below, became later known as the distinction between the *context of discovery* and the *context of justification*, and that is still being debated today.

---

he was – is that he did not focus on one area of science. I think Charles Pence (2018) is right when he diagnoses a profoundly lacking philosophical scholarship on Herschel today.

8 Agassi points out that Herschel's book was published in 1831 and not in 1830 as is often claimed: »title page to English edition says 1830, but the page with the portrait contains correct date« (Agassi 1969, 1, footnote 1).

9 In particular, Herschel discusses the example of Ampère's theory of electro-magnetism (1831, 202f., §§ 214 & 215). See Losee 2001, 104f.

2.2.2   *William Whewell: Introducing »Scientist« and Discoverer's Induction*

Among the first to cast a critical eye on Herschel's philosophy of scientific discovery was William Whewell (1794–1866). Whewell had befriended Herschel in their study days at Cambridge, and in late 1812 they had started to meet regularly on Sundays after chapel service, together with their friends Babbage and the economist Richard Jones (1790–1855), for what they called »philosophical breakfasts«.[10] Like Herschel, Whewell was a leading figure in his day. Unlike Herschel, however, Whewell, the son of a carpenter, came from a poor family and could start his career at Trinity College, Cambridge, only because of his prodigious mathematical abilities and because he received one of the few poor student scholarships – like Newton, Whewell was a so-called subsizar (Whitrow 1989, 71, Gibson 2019, 41f.). Nevertheless, Whewell had a lifelong association with Cambridge where he eventually became a star. He graduated Second Wrangler in 1816, was elected Fellow of Trinity in 1817, and Fellow of the Royal Society in 1820. Whewell served as Professor of Mineralogy at Cambridge, later as Professor of Moral Philosophy, he was the University's Vice Chancellor and also Master of Trinity – »some would say the most powerful position in the entire academic world« (Snyder 2011, 1).

Whewell wrote on a broad spectrum of topics, ranging from mathematics and crystallography to the study of the tides, and from Kant's work and moral theology to the philosophy and history of scientific discovery. In addition, he translated Plato's dialogues and Goethe's poetry. Whewell's variety of interests was so widely known that the essayist Sydney Smith said of him: »science is his forte and omniscience is his foible« (Blackburn 2016, 504), and his friend John Herschel:

> »a more wonderful variety and amount of knowledge in almost every department of human inquiry was perhaps never in the same interval of time accumulated by any man.« (Herschel 1868, liii, quoted in Snyder 2006, 1)

Besides, Whewell had a special talent for inventing neologisms; among others, he invented not only the terms »anode«, »cathode«, and »ion« (for Michael Faraday), but also the terms »physicist« and even »linguistics«.

---

10   Inspired by Bacon's writings, the four shared a common interest, namely the professionalization of science, generally speaking. In her both instructive and entertaining book *The Philosophical Breakfast Club* (2011), Laura Snyder investigates this more closely and provides a view from history of science combined with further biographical background information on the four protagonists. In addition, Snyder 2019 has been an important resource for the following presentation.

On occasion of the third meeting of the British Association for the Advancement of Science held at Cambridge on June 24, 1833, Whewell gave the opening lecture. »As the applause died down«, Laura Snyder tells us, »one man rose imperiously« (Snyder 2011, 2), and everyone was surprised to realize it was the celebrated poet Samuel Taylor Coleridge (1772–1834) who was also known to not have left his house in years for several reasons. Yet, Coleridge had felt obliged to make the long journey to his alma mater for the meeting. What Coleridge said shocked the members of the science association, namely that they should stop calling themselves »natural philosophers«, which was the most commonly used term for practitioners of science at the time.[11] Coleridge, however, clearly regarding himself as a ›true philosopher‹, pondering the cosmos from his armchair, found that »[m]en digging in fossil pits, or performing experiments with electrical apparatus, hardly fit the definition« – rather, for Coleridge, they were »practical men, with dirty hands at that« (Snyder 2011, 2). This was clearly an offense, and one which would have far-reaching consequences:

> »Whewell rose again, and quieted the crowd. He courteously agreed with the ›distinguished gentleman‹ that a satisfactory term with which to describe the members of the association was wanting. If ›philosopher‹ is taken to be ›too wide and lofty a term,‹ then, Whewell suggested, ›by analogy with *artist*, we may form *scientist*.‹« (Snyder 2011, 2–3)

With regard to the scientific achievements of Herschel and Whewell as well as to their philosophical reflections on science and discovery, it seems fitting that Whewell invented the term »scientist«, which helped to transform the natural philosopher into the modern scientist and which is still in use today.

Even though Whewell's general view of science was, like Herschel's, inspired by Bacon,[12] there are some differences between Herschel's and Whewell's views. In his book on natural philosophy, Herschel had claimed that we can arrive at laws in three ways:

(a) »[b]y inductive reasoning«,
(b) »[b]y forming at once a bold hypothesis«, or
(c) »[b]y a process partaking of both these« (Herschel 1831, 198–199, § 210).

---

11   Other terms were »savants« and »men of science«. For further discussion, also on the evolution of the term »science«, see Ross 1962.

12   Whewell even tellingly named the third edition of his *Philosophy of the Inductive Sciences, founded upon their History* after Bacon's work: *Novum Organon Renovatum* (Whewell 1858b).

As mentioned above, Whewell was the first to review Herschel's book on natural philosophy, and he argued, against Herschel, that verification is impossible when a hypothesis has been formed non-inductively (Whewell 1831, 400f.), i.e., he disagreed with Herschel's (b) and (c). In his book review and also in his later writings, Whewell, objected generally to a view according to which an essential part of science – making discoveries – consisted in what he regarded as unorganized or random guessing (Whewell 1849, Whewell 1858a, Whewell 1860). In short, Whewell did not believe that scientific discoveries were typically made by accident – not even Newton's famous discovery of the Law of Universal Gravitation.[13] Now, what began with a review of Herschel's book on natural philosophy brought Whewell later to develop a unique inductive methodology which he called »discoverer's induction«[14] and which Snyder briefly summarizes as follows:

> »Whewell followed Bacon in rejecting the standard, overly-narrow notion of induction that holds induction to be merely simple enumeration of instances. Rather, Whewell explained that, in induction, ›there is a New Element added to the combination [of instances] by the very act of thought by which they were combined‹ ([Whewell] 1847, II, 48). This ›act of thought‹ is a process Whewell called ›colligation.‹ Colligation, according to Whewell, is the mental operation of bringing together a number of empirical facts by ›superinducing‹ upon them a conception which unites the facts and renders them capable of being expressed by a general law. The conception thus provides the ›true bond of Unity by which the phenomena are held together‹ ([Whewell] 1847, II, 46), by providing a property shared by the known members of a class (in the case of causal laws, the colligating property is that of sharing the same cause).« (Snyder 2019)

2.2.3 *The Debate between William Whewell and John Stuart Mill*

Whewell's philosophy of discovery was prominently attacked by the English philosopher and economist John Stuart Mill (1806–1873), especially in his *System of Logic* ([1843] 1974). Having read both Herschel's and Whewell's books (Macleod 2020), Mill argued, against Whewell, that observation and induction are sufficient for arriving at facts about the world and for obtaining scientific concepts (Snyder 2006). By contrast, Whewell, as can be seen in the quote above, claimed that scientific reasoning involved, prior to the discovery of laws, an *a priori* element of concept formation. Without going much into detail here, it should be noted that Whewell's and Mill's disagreement on

---

13   We will return to this topic in § 2.4.4.
14   See Snyder 1997 and 2006 for detailed discussion.

science is only one facet of a wide-ranging debate between the two that ensued. Whewell had already read Mill's book soon after its publication; he wrote in a letter to Herschel on April 8, 1843:

> »[There is] a new book by young Mill about the philosophy of science, suggested in a great degree by your book on the same subject and by mine. There is in new books of this kind a satisfaction in which both you and I may have a share. I mean that notions and expressions, which were new and strange when we began to write, are now familiarly referred to as part of the uncontested truth of the matter. Mill agrees with you more than with me in the parts where we differ, but he does not appear to me an ally to set much store by; for, though acute and able, *he is ignorant of science* [...].« (Whewell to Herschel, Todhunter [1876] 2011, 315, my emphasis)

Macleod (2020) seems to agree with Whewell here when he calls into question whether Mill would have taken his line of argument if he had been better acquainted with both the history of science and with actual scientific practice – as both Whewell and Herschel undoubtedly were. If Whewell and Macleod are right in this, however, is hard to tell; plus, as Snyder reminds us in her remarkable book on the Whewell-Mill debate:

> »The differences between Mill and Whewell over induction in science cannot be understood without setting them in the broader context of their conflicts over the proper way to reform society.« (Snyder 2006, 4)

The broader context Snyder has in mind contributed *cum grano salis* to the fact that, from today's perspective, Herschel's and Whewell's views on science have unfortunately been overshadowed by Mill's works. What we can still learn from this, however, is that there was an intense and fruitful debate about science and the practice of discoveries at the time, leading not only to, as Whewell described it, our being familiar with new notions and expressions, but even to a fundamental influence on the achievements of actual scientific practitioners – perhaps most notably on those of Charles Darwin.

2.2.4   *John Herschel's Influence on Charles Darwin*

In 1831, his last year at Cambridge, Darwin read Herschel's book on natural philosophy which had just been published. Both this book – which »stirred up in [him] a burning zeal to add [...] to the noble structure of Natural Science« (Barlow 1958, 68) – and the meeting with Herschel at the Cape in 1836 had a considerable and continuing impact on Darwin's philosophy and attitude to research. In his autobiography, Darwin tells us:

»I felt a high reverence for Sir J. Herschel, and was delighted to dine with him at his charming house at the Cape of Good Hope, and afterwards at his London house. I saw him, also, on a few other occasions. He never talked much, but every word which he uttered was worth listening to.« (Barlow 1958, 107)

When it comes to his famous *Origin of Species* (Darwin 1859), the foundational work of evolutionary biology, it is remarkable to find that the influence of the philosophy of discovery of the time is already visible in the very first pages of the book: Darwin honors Whewell with a motto from the *Bridgewater Treatises* on the front page, even before a motto from Bacon.[15] In addition, Darwin praises John Herschel as »one of our greatest philosophers« in the second sentence of the introduction.[16] And as Pence's closer analysis of Darwin's overall argument reveals:

»Darwin's argument [unfolds] in precisely the way that we would expect given a desire to hold oneself to Herschel's methodological canons. Darwin begins by proposing a speculative hypothesis, grounded on an extensive analogical basis. He then sequentially follows Herschel's steps for the verification of that hypothesis, first demonstrating its adequacy and then its ability to account for a wide variety of phenomena which it was not originally proposed to explain.« (Pence 2018, 135)

There is an ongoing debate on the question of how much Herschel's and Whewell's philosophies have influenced Darwin's thought and his theory of evolution.[17]

What is more important in our context, however, is to see that the first half of the 19th century witnessed a flowering period of the philosophy of scientific discovery in which new concepts and debates emerged, leading not only to a better general understanding of scientific practice, but also contributing to actual scientific progress.

---

15   See Schweber 1989 for further details. I am particularly grateful to two highly supportive colleagues from biology, Paavo Bergmann (Konstanz) and Theo Michael Schmitt (Greifswald), for valuable discussions and additional information on Darwin's life and work.

16   The influence Herschel and Whewell had on Darwin has been worked out in greater detail, among others, by Michael Ruse and Paul Thagard, see Ruse 1975, Ruse 1978, and Thagard 1977. Cf. also Warner 2009.

17   In particular, David Hull and Michael Ruse, disagreeing in several respects, published a series of papers and books on that topic in the 1970s. Ruse (2010) and, in great detail, Honenberger (2018) offer an interesting explanation of these differences.

2.2.5   *After the 1850s: Philosophy of Scientific Discovery on the Decline*

Oddly, the situation changed, as far as I can see, around the middle of the 19th century. Well, not surprisingly, the topic of discovery still appeared in several places and it was also not insignificant. As a matter of fact, here are some examples: The notion of discovery is present in a fragment of Gottlob Frege ([1879–1891] 1983, 2–3)[18] – in this fragment, however, Frege only wants to set apart logic from other fields. Moreover, the topic of discovery is present in Ernst Mach's and Pierre Duhem's writings (Mach 1896, Mach [1905] 1917, Duhem [1908] 1998) – but actually it is somehow hidden in their remarks on analogy. Discovery also played a role in discussions of the Marburg school of Neo-Kantianism (Hermann Cohen, Paul Natorp, etc.) – their focus, however, was something else, namely the development of a so-called transcendental method which, in turn, committed them to a certain view of scientific progress (see Heis 2018). And even Henri Poincaré (1904) dealt with methods of discovery – he did so, however, only in the context of the question which role mathematical techniques can play in the generalization of empirical findings (see Meheus 1999).

Against the background of examples like these, we can concede that the topic of scientific discovery was only somehow present in these and other works after the 1850s. Yet, we also see that there was no greater systematic study or philosophical debate focussing on the concept of scientific discovery. In other words: After the 1850s, the philosophy of scientific discovery was on the decline.

2.2.6   *Hans Reichenbach, Karl Popper, and the Context Distinction*

This seemed to change again in the 1930s, especially with the publication of a popular book by Karl Popper (1934), entitled in English »*The Logic of Scientific Discovery*«. This title, however, is nothing but a misleading translation of the German original (»*Logik der Forschung*«, which should rather be translated as »*The Logic of Scientific Research*«). Anyway, in his book, Popper made use of a certain distinction which was crucial for his overall argument (Popper 1934, chap. 1, sec. 2) and which he ascribed in a short article only one year later to Hans Reichenbach (Popper 1935). The distinction Popper referred to, however, was not original with Reichenbach, and we have already encountered it in connection with John Herschel's philosophy of discovery: We are talking about the distinction between the context of discovery and the context of justification (see § 2.2.1). What is correct, though, is that, in his book *Experiment and Prediction* ([1938] 2006), Reichenbach re-introduced this distinction in a modern logicist version. Like Herschel, Reichenbach regarded actual processes of

---

18   Thanks to Stefan Reins (Göttingen) for pointing this out to me.

theory generation as irrelevant to their justification. Unlike Herschel, however, Reichenbach and Popper argued that philosophy should be concerned with the context of justification – and not at all with the context of discovery, which in their view should rather be dealt with in psychology. One of Popper's most remarkable passages on this issue reads as follows:

> »The initial stage, the act of conceiving or inventing a theory, seems to me nei-ther to call for logical analysis nor to be susceptible of it. The question how it happens that a new idea occurs to a man – whether it is a musical theme, a dramatic conflict, or a scientific theory – may be of great interest to empirical psychology; but it is irrelevant to the logical analysis of scientific knowledge. This latter is concerned not with *questions of fact* [...], but only with questions of *justification or validity* [...]. [...] [T]here is no such thing as a logical method of having new ideas, or a logical reconstruction of this process.« (Popper 1959, 31f.)

It is this exclusion of the context of discovery that Kitcher had in mind in the passage I quoted in the beginning of this section (see § 2.1 above), writing that discovery »fell out of favor as an influential group of philosophers argued that there was no general method of scientific discovery« (Kitcher 2001, 109).[19] And Kitcher is right: The context of discovery was lastingly removed from philoso-phy and relegated to psychology. What is more, the distinction between the two contexts which enabled this relegation in the first place was not even origi-nal with these thinkers.

2.2.7 *Norwood Russell Hanson's Underrated Logic of Discovery*

While the English title of Popper's book is misleading, this is not the case with Norwood Russell Hanson's 1958 book *Patterns of Discovery* (Hanson 1958a). Moreover, in contrast to Popper, Hanson claimed that there is indeed a logic of discovery, i.e., he explicitly included the context of discovery in the domain of philosophical research. Building on the so-called context distinction, Hanson argued that »in some cases one's reasons for entertaining an [sic] hypothesis and his reasons for accepting it are logically different« (Hanson 1958, 1077). Hanson was, as this already suggests, concerned with the discovery of hypoth-eses and theories, in particular. With regard to his project, however, Hanson was seemingly the only one who was really convinced of it, so that it was, in the

---

19   Speaking of »*an* influential group«, Kitcher seems not to distinguish between the move-ments of Logical Positivism and Logical Empiricism nor between the two correspond-ing main groups active at the time, the Berlin Group and the Vienna Circle. Kitcher is, however, right that the distinction in question has to do with both of these groups: Reichenbach was a leading figure and founder of the Berlin Group and Popper was in close contact with the Vienna Circle without being a member of it (see Naraniecki 2010 for further details).

end, »a one-man campaign for the logic of discovery« (Lund 2010, 104). Thus, even though Hanson had what I regard a promising approach to the philosophy of scientific discovery, it did not spread or receive adequate attention – unfortunately, Hanson's work is today mostly associated only with the idea that observation is theory-laden.

2.2.8   *The Philosophy of Scientific Discovery Today*

Much more could be said on the further development of the philosophy of scientific discovery, and not only about the role of other prominent philosophers of science, such as Paul Feyerabend, Thomas Kuhn, or Imre Lakatos. However, I leave it at that here, claiming that the concept of scientific discovery has not at all received due recognition from a systematic point of view.

And I leave it at that not only because of the aforementioned evaluations of the current status of the philosophy of scientific discovery by Kitcher and others that fit my own evaluation, but also in the light of a remarkable entry in what is likely to be regarded the best philosophy of science encyclopedia currently on the market. I am talking about the roughly 1000-pages 2-volume Routledge encyclopedia *The Philosophy of Science* (Sarkar & Pfeifer 2006). The publisher's book description is certainly correct in stating that the encyclopedia »brings together an international team of leading scholars to provide over 130 entries on the essential concepts in the philosophy of science«.[20] But then, what does this encyclopedia say on the concept of scientific discovery? Let me cite the *complete* encyclopedia entry here:

> »SCIENTIFIC DISCOVERY: *See* Scientific Change«
> (Sarkar & Pfeifer 2006, 732)

Yes, that's the complete entry. In other words: Even though the concept of scientific discovery *should* clearly be an essential concept in the philosophy of science, there *is really no* entry on it in the best current philosophy of science encyclopedia! And, by the way, the entry on scientific change is not really about discovery – rather, it is ... well, about scientific change.

Now, what can we learn from all this? At least, we can learn how to answer the question we posed at the beginning of this section: Don't we have a philosophy of scientific discovery at hand already? Answer: No, there is currently no sustained and intensive systematic philosophical research on the topic of scientific discovery. The last time we had a serious philosophy of scientific discovery was around the middle of the 19th century.

---

20   www.routledge.com/The-Philosophy-of-Science-2-Volume-Set-An-Encyclopedia/
      Sarkar-Pfeifer/p/book/9780415939270.

2.3 The Need for a Philosophy of Scientific Discovery

We have seen that, curiously, the topic of scientific discovery has received hardly any attention in systematic philosophy for over 150 years. But, then, do we need a philosophy of scientific discovery at all? I present three arguments that we do: the no discoveries argument (§ 2.3.1), the discovery of the concept of discovery argument (§ 2.3.2), and the Grand Challenges argument (§ 2.3.3).

2.3.1 *The No Discoveries Argument*

My first argument why we need a philosophy of scientific discovery is based on a simple thought experiment which, basically, amounts to the following question: What if there were no discoveries made anymore in a certain scientific discipline?

Imagine the following scenario: In a not too far future, the world develops in such a way that no new discoveries are made in, say, the field of physics.[21] Let us assume that some physicists notice this fairly early, but are not seriously concerned at first. It is only after some time that more and more questions appear and also the (public) pressure on the people in physics rises. Let us further assume that, finally, in our imaginary scenario, the world's leading physicists, after thorough examination, publicly concede and announce that there is really nothing to discover in physics anymore.[22] We would surely be surprised by such a development – but what should and what would we do? For instance, should the financial support for research in physics be maintained? Should we continue to employ physicists? Would the departments of physics still look the same? If not, how should they change? Would physics still be part of our universities? Would we still think of physics as a scientific discipline?

This motivates the following argument: The concepts of science and discovery are so closely intertwined that we can neither easily nor decisively answer any of the aforementioned questions. But if the concept of science is so closely tied to the concept of discovery, we need a philosophy of scientific discovery in order to get a better understanding of science.

---

21   In case you are unhappy with this example, just pick a different one.

22   We all know that there is usually no consensus on the question of who we should regard as belonging to the ›leading‹ persons in a field. This, however, should not present a problem with regard to our imaginary scenario which can easily be modified (see Michel 2011, Michel 2013). Instead of speaking of the world's leading physicists, we could, for instance, assume that the bad news is delivered by, say, the world president of physics or even by a physically omniscient demon (see Michel 2020a).

2.3.2 *The Discovery of the Concept of Discovery Argument*

My second argument why we need a philosophy of scientific discovery does not build on thought experimentation, but rather on a mixture of insights from the history of science, etymology, and the history of ideas. Still, in spirit, it is similar to the first argument in so far as it is meant to demonstrate the close connection of the concepts of science and discovery. More precisely, the argument aims to show that the concept of (modern) science builds on the concept of discovery. So, we should regard the discovery of the concept of discovery as indispensable for the development and understanding of (modern) science.

Let us begin with the following observation: It is a matter of controversy among historians of science when exactly the so-called scientific revolution,[23] which marked the emergence of modern science,[24] took place. While some hold that it began in 1543, with the publication of Nicolaus Copernicus' *De revolutionibus orbium coelestium*, others claim that it did not begin before 1572, when Tycho Brahe observed a supernova. And as regards the completion of the scientific revolution, some argue that it was in 1687, with the publication of Newton's *Principia*, while others think that it was only in 1704, with the publication of Newton's *Opticks*. What is not a matter of controversy, however, is that, since its very beginnings, modern science has been permeated with discoveries: from the anatomical discovery of the smallest bone in the human body (the stapes)[25] to Galileo's astronomical discovery of Jupiter's four moons[26] and the case of the chemical discovery of oxygen,[27] to name just a few

---

23   The term »scientific revolution« was introduced by Alexandre Koyré in the 1930s and popularized, among others, by Herbert Butterfield and C.P. Snow in the 1940s and 1950s. For further details see Wootton 2016, chap. 2.

24   Wootton (2016) speaks of the »invention« of science – it is also the title of his book. Interestingly, Weinberg (2015, xi) tells us that he thought of using the same (sub)title for his book, but that, eventually, he decided to speak of the »discovery« of science. For several reasons, I think both solutions are not uncontroversial.

25   While for instance Dispenza, Cappello, Kulamarva & De Stefano (2013) defend the common view that Giovanni Filippo Ingrassia (1510–1580) should be credited with the discovery of the stapes, Mudry (2013) argues that this is not as clear as it seems.

26   According to Livio, Galileo's discovery »was not only of historical significance – these were the first new bodies revealed in the solar system since antiquity – it also demolished one of the serious objections to the heliocentric model« (Livio 2020, 68). And apparently, Galileo realized at once how important his discovery was and »stopped writing his observation notes in Italian and started writing them in Latin – he was planning to publish« (Wootton 2016, 86). We will return to the topic of publication and its crucial role within scientific discovery processes (see § 2.5.2). Thanks to Mario Livio for discussion.

27   The discovery of oxygen seems to be more complex than some other discovery cases. It is, for instance, unclear who discovered it (Scheele, Priestley, or Lavoisier?) and it is unclear when exactly it was discovered (between 1772 and 1812, see Wootton 2016, 87).

famous examples from the earlier days of modern science. But once we take a closer look at etymology and the history of ideas, it is no surprise that discoveries play such a crucial role in modern science.

From David Wootton's excellent etymological overview we learn that, curiously, there was no word for discovery available in European languages until the time of Christopher Columbus' expedition which led to the so-called discovery of America (Wootton 2016, Wootton 2017). There was only a word for discovery in Portuguese at the time (»discobrir«), and it only appeared around 1485 (Wootton 2016, 57). The new word, however, began to spread across Europe by the beginning of the 16th century; by the middle of the 16th century, it had already been translated into French, Italian, Spanish, German, and English.[28] But how was it used, what did it mean?

> »The core meaning of ›discovery‹, after 1492, is not just an uncovering or a finding out: someone who announces a discovery is, like Columbus, claiming to have got there first, and to have opened the way for all those who will follow. [...] Discoveries are moments in an [sic] historical process that is intended to be irreversible. The concept of discovery brings with it a new sense of time as linear rather than cyclical. If the discovery of America was a happy accident, it gave rise to another even more remarkable accident – the discovery of discovery.« (Wootton 2016, 60–61)

What Wootton has in mind here, or so I read him, is that this linguistic development or change is an indication of the emergence of a new concept: the concept of discovery. Before we had this concept, »the greatest achievements of civilization were believed to lie not in the present or the future but in the past, in ancient Greece and classical Rome« (Wootton 2016, 61). The discovery of the concept of discovery, therefore, »transformed the world, for it has made modern science [...] possible« (ibid., 62) – and simply because the idea of discovery »is not itself a scientific idea but rather an idea that is *foundational for science*« (ibid., 103, my emphasis).

What does this mean for our argument? Following Wootton, we see that the discovery of the concept of discovery constitutes a necessary condition for the emergence of modern science. In other words: Science as we know it would not have been possible without the (discovery of the) concept of discovery. But if the concept of science is based on the concept of discovery in this way, we need a philosophy of scientific discovery in order to get a better understanding of modern science.

---

28   Clearly, the invention of the printing press and with it the so-called printing revolution played an important role, see Wootton 2016, *passim*.

2.3.3   *The Grand Challenges Argument*

Unlike the two preceding arguments, my third argument does not primarily focus on a better theoretical grasp of what science is, but rather on the broader question of how to approach the so-called Grand Challenges. Roughly speaking, the term »Grand Challenges« is used both to identify and to refer to the most important global problems of the 21st century that we can plausibly address through coordinated and collaborative effort by means of science, technology, and innovation. The general idea behind the formulation of the Grand Challenges is inspired by David Hilbert's articulation of the 23 most important unsolved mathematical problems in 1900 which he hoped would enable further progress in the field (Hilbert 1900). Hilbert's story gained momentum when Bill Gates announced his so-called Grand Challenges in Global Health initiative in 2003:

> »When launched, Grand Challenges in Global Health committed $481.6 million for projects focusing on 14 major global health challenges. The aim was to engage creative minds across scientific disciplines to work on solutions that could lead to breakthrough advances for those in the developing world.«[29]

Since then, the Grand Challenges movement has grown steadily and now comprises a whole »family of initiatives fostering innovation to solve key global health and development problems«.[30] According to the European Commission's »Horizon 2020« program, these include the following:

> »1. Health, demographic change and wellbeing
> 2. Food security, sustainable agriculture and forestry, marine, maritime and inland water research, and the Bioeconomy
> 3. Secure, clean and efficient energy
> 4. Smart, green and integrated transport
> 5. Climate action, environment, resource efficiency and raw materials
> 6. Inclusive, innovative and reflective societies
> 7. Secure & innovative societies«[31]

And according to the »Strategy for American Innovation«, first issued by the administration of US President Barack Obama,

---

29   See https://gcgh.grandchallenges.org/about. As Gates remarked in October 2020, »ending the current pandemic« is clearly one of the Grand Challenges, see https://grandchallenges.org/video/keynote-remarks-bill-gates.

30   https://grandchallenges.org/about.

31   https://ec.europa.eu/programmes/horizon2020/sites/horizon2020/files/InfoKit_UK_240214_Final.pdf.

> »Grand Challenges [...] can have a major impact in domains such as health, energy, sustainability, education, economic opportunity, national security, and human exploration. Also [...] the question ›what should we do‹ is arguably as or more important than ›what can we do.‹ This is not just a technical question, it is a question that relies on imagination, creativity, values, as well as individual and shared views on how we define progress.«[32]

I cannot go into detail here, but I think the core idea of the Grand Challenges initiative should be clear: We are talking about the incredibly ambitious and complex task of identifying and also meeting the greatest challenges we humans are currently facing. Even though the Grand Challenges initiative comes with a variety of controversies, its general significance cannot be overestimated.

What is important for our purpose here has been nicely illustrated by Alan Liu, a professor in the English Department at UC Santa Barbara, in a somewhat different context in a blog post. Liu writes that

> »we use words like *invention, innovation*, and *breakthrough* to describe the most hopeful visions for the future of humanity. We pin our hopes on technological and other breakthroughs [...]. We even have a name for the greatest human challenges [...]. We call these ›grand challenges.‹ [...] Yet not one of the words *invention, innovation*, and *breakthrough* are as powerful as the word that encompasses them all and gives them their full human meaning. That word is *discovery* [...].« (Liu 2012)

With this in mind, we arrive at the following argument: If we want to meet the Grand Challenges, we have to both identify and understand their core aims – which, in turn, means that we have to reflect upon their basic concepts. As their various formulations indicate, the Grand Challenges can only be met with the help of science, especially by means of invention, innovation, and breakthrough. Inspired by Liu, it seems plausible to claim that concepts like invention, innovation, and breakthrough are powered by the concept of scientific discovery – the concept »that encompasses them all«. But then, since it is crucial for the future of humanity that we meet the Grand Challenges, it seems plausible that we also need to reflect upon the concept of scientific discovery. In short: We need a philosophy of scientific discovery, and we need it urgently.

---

32   https://obamawhitehouse.archives.gov/sites/default/files/strategy_for_american_innovation_october_2015.pdf, see page 85.

2.4   Different Cases of Discovery

To sum up the argument of the last two paragraphs, the situation we find our-selves in is this: We need a philosophy of scientific discovery for several rea-sons, but there hasn't been one worthy of that name for over 150 years now. That is why I think it is high time to change the situation and move toward a philosophy of scientific discovery.

To begin with, I take a closer look at the concept of discovery by considering the following three discovery cases: Albert's discovery (§ 2.4.1), Blackbeard's discovery (§ 2.4.2), and Charles' discovery (§ 2.4.3).[33] After discussing these more or less fictional cases, I also consider the importance of real discovery case studies (§ 2.4.4).

2.4.1   *Albert's Discovery*

Let us begin with a simple everyday example: Upon slipping his foot into his sock, Albert discovers a hole in it. Even though we would probably not want to consider this to be a case of *scientific* discovery, it clearly is a discovery case and it is also advisable to take a closer look at it.

Among others, we may derive the following: The hole in the sock (the object of the discovery) obviously came into existence before Albert (the subject of the discovery) discovered it. Moreover, if Albert had known about the hole beforehand, it would not have been a discovery for him. Thus, it is a necessary precondition for Albert's discovery that he had no prior knowledge of the hole. This, in turn, points to a change in Albert's beliefs, for he now believes that there is a hole in his sock – something he did not believe before. Generally, instead of speaking of a discovery here, we could just as well say that Albert »found, saw, or stumbled across the hole in the sock« (Michel 2019, 427).

2.4.2   *Blackbeard's Discovery*

My second discovery case is similar to Albert's case in that we would not con-sider it to be a *scientific* discovery case. Unlike Albert's case, however, it does not involve an everyday scenario, but one with which we are acquainted from fiction. Here it goes (see also Michel 2019, 427f.): Imagine that, under dubi-ous circumstances, Blackbeard gets hold of an old treasure map hidden in an empty old bottle of rum. Convinced that the map is no fake, he gathers a crew of clandestine men and sets sail to the alleged treasure island. After some days in rough waters, Blackbeard and his crew arrive at the island and find the place

---

33   The following presentation builds on the line of argument in Michel 2019 in which I spell out these three cases in more detail (pp. 426–428).

indicated by a black cross on the map. Blackbeard commands his men to start digging in the sand. Soon they find a treasure chest full of gold and diamonds.

Blackbeard's discovery may not be a scientific one, but it certainly is a discovery. It is also different from Albert's discovery in several respects. For one thing, it is unclear in Blackbeard's case who exactly the subject of the discovery is: Is it just Blackbeard? Or is it also his crew? Moreover, in contrast to Albert's case, it was neither serendipitous nor surprising for Blackbeard to discover the treasure;[34] Blackbeard had prior knowledge (belief, expectation) with regard to the object of the discovery, i.e., the treasure. At least, he expected the treasure to be there, whatever it exactly was or contained. In contrast to the case of Albert who did not even plan his sock hole discovery, this is even a necessary precondition for Blackbeard's discovery – for if Blackbeard didn't have any belief or expectation about the treasure, he would certainly not have started the whole enterprise.[35]

2.4.3 *Charles' Discovery*

Let us contrast Albert's and Blackbeard's non-scientific discoveries with a scientific one – and since we have to start somewhere, let us begin with a biological discovery (see Michel 2019, 428f.). Assume that Charles, after years of intensive study, has acquired an enormous body of biological background knowledge and is now a recognized biological expert on fish. As I have put it elsewhere:

> »Charles walks along the beach and suddenly sees, in a rock pool, some beings he has never seen before. Even a layperson, would probably have been able to describe the beings in question as fish (i.e., in everyday or prescientific language) that float in the rock pool, but it is only because of Charles' biological background knowledge and his being an expert on (the paraphyletic group of) fish that he has the strong and justified conviction that these individuals belong to a species that has not yet been captured by any biological taxonomy yet. In order to rectify this situation, Charles catches some of the fish and takes them to a laboratory. His closer investigation of the exemplars leads to a solidification of his original conviction. Charles is a leading expert in his field of research and he does not want to make a mistake or overlook anything. So, he thoroughly checks the textbooks and taxonomies again, and he also has longer discussions with several of his colleagues about certain details of his findings. After a couple of weeks, Charles is convinced and prepares a manuscript for publication in which he gives a detailed description of his findings and also proposes a new species name. By submitting the manuscript to a respected biological journal, Charles declares his findings as new biological species discovery. Several weeks later, the

---

34   We will turn to the topic of serendipity below, § 2.4.4.
35   For further details see Michel 2019, 428.

peer-review process including Charles' revisions of his paper is over and Charles' paper is published. This is a case of success: The scientific community accepts Charles' proposal, his findings count as a species discovery new to biology and Charles counts as the discoverer of this species.« (Michel 2019, 428)

Comparing Charles' discovery with Albert's and Blackbeard's, a couple of interesting features with regard to the concept of discovery stand out. To begin with, it is not a necessary precondition – as it is in Albert's case – that Charles had no prior knowledge of the exemplars of fish in question. Even if he had known about the exemplars before, he could still have become the discoverer (or co-discoverer) of the new biological species – for instance, if someone else had found the exemplars and asked Charles as an expert for support. Another difference to Albert's case is that Charles' discovery is not a case of pure coincidence[36] – even though some parts of Charles' overall discovery process happened accidentally, especially his finding of the exemplars. For Charles' discovery, a prior body of knowledge (both knowledge-that and knowledge-how)[37] was necessary as was the finding of the exemplars and also the collective acceptance (recognition, acknowledgment)[38] of this finding as a new discovery by a relevant scientific community. In addition, we should not overlook an important aspect Charles' story entails, namely that, once the findings are accepted as a new biological species discovery, we have gained new knowledge in biology. This new knowledge includes, among others, that we now know that individuals with certain characteristics (further specified by Charles) – individuals we would simply refer to as »fish« in everyday language – belong to a certain biological species (named so-and-so by Charles). To put it in a nutshell, we should regard Charles' biological discovery as comprising three core structural features: *finding, acceptance*, and *knowledge*. Moreover, I assume that these three features are, quite generally, indispensable for any scientific discovery.

2.4.4   *The Importance of Real Discovery Case Studies*

Before we turn to the three core features in greater detail, let us remind ourselves that the three discovery cases we have discussed so far comprise more

---

36   The same holds for Blackbeard's discovery.

37   We will turn to this distinction below, § 2.5.3.

38   We will turn to the topic of acceptance below, § 2.5.2. In order to prevent misunderstanding, let me point out that my reading of acceptance does not imply approval. I agree with Searle here: »Acceptance, as I construe it, goes all the way from enthusiastic endorsement to grudging acknowledgment, even the acknowledgment that one is simply helpless to do anything about, or reject, the institutions in which one finds oneself« (Searle 2010, 8).

or less fictional scenarios. Is that a problem? No, it isn't – quite the contrary, we can learn a lot from analyzing fictional cases (see Michel 2011, Michel 2013, Michel 2020a). Yet, it is also crucially important to supplement our discussion of fictional discovery cases with analyses of real scientific discovery cases in order to get a better understanding of what it means to make discoveries in science. To illustrate what I mean by this, let us take a brief look at some real cases and see what we can learn from these.

Thinking of real discoveries, several examples easily come to mind. Some of these are iconic cases, such as Archimedes' discovery that the buoyant force on an object submerged in a fluid is equal to the weight of the fluid which is displaced by that object. Archimedes, who lived in the third century BCE in Syracuse on the island of Sicily, was probably antiquity's most famous mathematician (see Weinberg 2015, 37–39).[39] As legend has it, Archimedes recognized the significance of what we call Archimedes' Principle today when he was getting into a bathtub, observing that the more his body sank into it the more water ran out over the tub. We are told that Archimedes immediately jumped out of the tub and rushed home naked through the streets of Syracuse, shouting »Eureka, eureka!« – which can be translated as »I have found it, I have found it!«.

It may be a moot point whether this story is true in all its curious details (probably not); what is usually not disputed, however, is *that* Archimedes really made this discovery. But what is it that makes it an iconic case? First of all, the focus of the story is not so much the object of the discovery (Archimedes' Principle), but rather its subject – more precisely: the instant of Archimedes' finding. What the story illustrates is how the solution to the puzzle Archimedes wanted to solve, is, all of a sudden, obvious and clear to him, leading to his spontaneous reaction and his exclamation of joy or satisfaction. Archimedes finally found what he was looking for, had an important insight and formed a new belief that became common knowledge. Moreover, Archimedes' story has also been linguistically formative, giving us commonly used terms like »eureka moment« and »eureka effect«.[40] So, what we can take from this discovery story is that Archimedes' eureka moment, his sudden insight, his finding plays a crucial role within the discovery process.

---

39   As becomes obvious in several places of Wootton's careful and detailed historical analysis (Wootton 2016), the fact that Archimedes made discoveries in the third century BCE already does not contradict or even refute the discovery of the concept of discovery argument (§ 2.3.2). Thanks to Michael Pohl (Rostock) for pointing this out.

40   Terms like »Aha! moment« and »epiphany« are related, but have different histories.

The story of Archimedes' discovery is driven by a powerful idea, the eureka idea, and it is hardly surprising to see that it is not unique with Archimedes' case. Rather, the idea is present in several stories in the history of science – just think of another iconic discovery case, the story of Newton and the apple. According to this story (true or not), his observation of the falling apple gave Newton, like Archimedes' observation of the overflowing water, a eureka moment which led to his discovery of the Law of Universal Gravitation.

Another iconic eureka story is that of Einstein's hitting on the core idea of his famous (Special) Theory of Relativity (Einstein 1905). As we know from his autobiography, Einstein had been thinking about the movement of light since the age of 16 already (Einstein 1949, 53). It was, however, only ten years later, in 1905, in his so-called *Annus mirabilis*, that Einstein came up with a thought experiment which involved not only the movement of light, but also his ideas on the question of what it means that two events happen simultaneously. In a nutshell (see Einstein [1917] 2009, 16–18), assume that two bolts of lightning strike the front and the back of a long and moving train and also assume that these two bolts of lightning are simultaneous with reference to (an observer on) the railway embankment. Einstein wondered whether the two bolts of lightning were also simultaneous with reference to (an observer on) the train. The short answer is: No. From this, Einstein concluded that a statement about the time of an event is meaningless unless the reference system that the statement refers to is also given. His thought experiment brought Einstein to have, like Archimedes and Newton, a eureka moment: All of a sudden, Einstein had hit on the core idea of Special Relativity.[41]

What the stories of Archimedes, Newton, and Einstein have in common is that they, via the eureka moments they contain, illustrate an important characteristic of the notion of finding: All of these stories focus on the instant in which an individual has a sudden insight, a certain psychologically fascinating kind of moment. Einstein's case, however, is different from the other two in that Einstein's eureka moment is not based on observation of something ›external‹ (overflowing water, falling apple), but rather on the employment of an ›internal‹ device (thought experiment) that requires, among others, certain creative, imaginative, and methodological skills. So, a closer look at these three iconic cases already allows us to see that some eureka moments are based (more) on observation while others are based (more) on thought experimentation.[42]

---

41   A case that is similar in this respect is that of Galileo refutation of Aristotle's claim by way of thought experimentation (see Brown & Fehige 2019 and Livio 2020, 30–36, but see also Schrenk 2004).

42   Obviously, there is much more to be said on this – and not only by way of more eureka stories. I am grateful to Bill Lycan who pointed out to me that the eureka topic is also closely connected to the topic of humor and joy (see Hurley, Dennett & Adams 2011).

Now, apart from these, there are more iconic cases of discovery, and not all of them are about eureka moments. A famous example is Columbus' so-called discovery of America.[43] Without going much into detail here, I would like to point out three aspects of this discovery case that usually do not get much attention: science funding, the role of serendipity, and making joint discoveries. Let us have a look at these.

First of all, Columbus' discovery would not have been possible without his sponsors Ferdinand of Aragon and Isabella of Castile – in other words: King Ferdinand and Queen Isabella's funding was a necessary precondition of Columbus' discovery. In this respect, Columbus' case clearly differs from the cases of Archimedes, Newton, and Einstein, for these cases were, more or less, independent of their financial background. It would, however, be completely wrong to conclude that the financial background is not important for making scientific discoveries. Quite the contrary, just think of the role that finance plays in »today's fast-paced, funding-starved, impact-driven atmosphere« (Livio 2022) in which we long for scientific discoveries and breakthroughs. Let me illustrate this with the help of one well-known and fairly recent example. On 8 October 2013, François Englert and Peter Higgs were jointly awarded the Nobel Prize in physics for

> »the theoretical discovery of a mechanism that contributes to our understanding of the origin of mass of subatomic particles, and which recently was confirmed through the discovery of the predicted fundamental particle, by the ATLAS and CMS experiments at CERN's Large Hadron Collider.«[44]

As the quote indicates, the discovery in question consists of two steps. Its first step began in 1964 with the independent publications of Englert (together with his colleague Robert Brout, now deceased) and Higgs, which led to the so-called Higgs mechanism (also known as the Brout-Englert-Higgs mechanism). In a second step, Brout's, Englert's, and Higgs' »theoretical discovery« was confirmed by the (practical) discovery of the so-called Higgs particle at the CERN laboratory for particle physics outside Geneva, Switzerland, on 4 July 2012. According to the press release of the Royal Swedish Academy of Sciences,

---

43   I speak of a »so-called discovery« in a distancing manner because there is controversy about whether it really was a discovery or not. What is less controversial, however, is that Columbus, through his four voyages to the region between 1492 and 1502, introduced the Americas to Western Europe, and what is usually agreed upon is that Columbus facilitated the influx of Europeans that, ultimately, led to the formation of new nations, including the USA, Canada, and Mexico. (For more details see e.g. Appleby 2013.) Anyway, whether Columbus' discovery really was one or not is irrelevant for my argument here.

44   https://www.nobelprize.org/prizes/physics/2013/summary/.

> »CERN's particle collider, LHC (Large Hadron Collider), is probably the largest and the most complex machine ever constructed by humans. Two research groups of some 3,000 scientists each, ATLAS and CMS, managed to extract the Higgs particle from billions of particle collisions in the LHC.«[45]

There are many fascinating questions surrounding this discovery. Since we are, however, focusing only on the topic of funding here, it is worth noting that it was one of the main motivations for the LHC to search for evidence of the so-called Higgs particle. Now, the LHC was not unexpensive, but »cost on the order of $10 billion and involves thousands of scientists from dozens of countries« (Greene 2013). In other words: As in Columbus' case, this discovery would clearly not have been possible without its financial background, and this is obviously also true of many other discoveries in science. Generally speaking, however, funding is a highly important, though not indispensable or essential aspect of making scientific discoveries.

Secondly, Columbus' discovery was a rather accidental one (in a way, similar to Albert's case, § 2.4.1). Columbus was looking for something different and was also wrong in several respects (see e.g. Eco 1999, 5–9). This, however, is not unique with Columbus' case; scientific folklore is filled with tales of accidental or chance discoveries, just think of the discoveries of the Dead Sea Scrolls (also known as the Qumran Caves Scrolls), polytetrafluoroethylene (better known as Teflon), penicillin, or X-rays – all discovered by accident. There is even a special term for this kind of discovery, a term originally coined by the English writer Horace Walpole in a letter to his friend Sir Horace Mann on 28 January 1754: Impressed by the Persian fairy tale »The three princes of Serendip«, in which the princes repeatedly, by accident and sagacity, make discoveries of things they were not looking for, Walpole introduced the term »serendipity« to describe some of his own accidental discoveries (Merton & Barber [1958] 2004). Since then, the word »serendipity« has been used to refer to the process of finding something interesting or valuable by chance – the word has become so popular that it was even voted Britain's favorite word in the year 2000.[46] As regards the role of serendipities in science, several further questions come to mind: What exactly counts as a serendipitous scientific discovery? Are there different kinds of serendipities? How important are they with regard to their respective overall discovery processes? Generally speaking, this is, apparently, a fascinating topic, partly also because serendipities – sometimes in connection with instructive blunders in science (Livio 2013) – have played different

---

45   https://www.nobelprize.org/uploads/2018/06/press-21.pdf.
46   http://news.bbc.co.uk/2/hi/uk_news/930319.stm.

key roles in many real scientific discovery cases. Therefore, we can say that serendipity is clearly an important topic to be dealt with within the philosophy of scientific discovery.[47]

Thirdly, similar to the case of Blackbeard's discovery (§ 2.4.2), it is unclear in Columbus' case who exactly the subject of the discovery was; the discovery is often (justified or not) attributed solely to Columbus, even though he clearly needed his crew for it, too. Now, as before, this is not unique with Columbus' case; rather, it is quite common that the subject of a scientific discovery is not only one person. Without going much into detail here, just think of other famous scientific discovery cases, such as the discovery of nuclear fission, which began in 1938 and involved Otto Hahn, Lise Meitner, and others, or the discovery of the structure of DNA in 1953 by James Watson, Francis Crick, and Rosalind Franklin. As a closer look at real discovery cases reveals, the subject of a scientific discovery may vary from individual researchers up to research teams comprising several people. Speaking of joint discoveries, it is advisable to analyze in every detail the different roles of the people involved – for instance, the roles that Fritz Straßmann and Otto Frisch played with regard to the discovery of nuclear fission or the very different roles that Linus Pauling and Maurice Wilkins played with regard to the discovery of the DNA structure. Doing so will probably not only lead to a better understanding of the discoveries in question, but also to certain re-evaluations in several respects, such as the scientific meritocracy and eponymy (»naming things after people«, Wootton 2016, 98) that accompany individual discovery cases.[48]

This brief overview was meant to illustrate that and how we can identify further features of scientific discoveries when we take a closer look at real discovery cases. We have seen that some discovery stories contain eureka moments, some of which are based on observation, others on thought experimentation. Sufficiently detailed real discovery case studies can also help to get a better understanding of the importance of science funding, serendipity and the roles of people involved in a given discovery process.[49] I am confident that real discovery case studies will not only enable us to identify even more discovery features, but also to distinguish different kinds of discoveries, such as what I

---

47   It's good to see that there is already some work on different aspects and questions connected to the topic of serendipity in science. See, for instance, Copeland 2017, Copeland 2018, Yaqub 2018, as well as the articles in de Rond & Morley 2010.

48   In my view, the discoveries of the planets Uranus and Neptune may be example cases.

49   This may also help to determine who is to be credited as (co-)discoverer(s).

call nested discoveries,[50] derived discoveries,[51] or complex discoveries.[52] Apart from these, I am sure that we can also learn from discovery case studies about the revisability,[53] the different degrees of (social) relevance,[54] and the temporal dimensions of scientific discoveries,[55] to name just a few.

2.5   The Structure of Scientific Discoveries

Finally, let us turn to the aforementioned crucial structural features of scientific discoveries – finding (§ 2.5.1), acceptance (§ 2.5.2), and knowledge (§ 2.5.3). I regard these, in contrast to the features discussed above (esp. in § 2.4.4), as indispensable features of scientific discovery processes. In what follows, I will lay out a program with regard to the philosophy of scientific discovery. Clearly,

---

50   Here is an example of a nested discovery: In 2020, a giant crystal was discovered in the Windloch cave, which was only discovered in 2019 (Ebenau & Voigt 2019, Brämer 2020). We may also make a related distinction between primary and secondary discoveries; thanks to Florian Lippert (Groningen) for pointing this out.

51   My distinction between derived and non-derived discoveries (as well as my related distinction between genuine and non-genuine discoveries) is meant to serve as a tool to (re-)assess discoveries especially in disciplines (or fields) that are complex in so far as they consist of interrelated subdisciplines (or subfields), such as environmental studies, medicine, or theology. Discoveries in church history, for instance, may serve as examples of derived (or non-genuine) discoveries in theology, but what about non-derived (or genuine) discoveries? Did Hartshorne even have a point here when he asked: »Did Anselm, in his ›Ontological Argument‹ [...] make one of the greatest intellectual discoveries of all time, or did he merely fall into an interesting blunder?« (Hartshorne 1965, 3). Are there other examples as well? See Michel & Göcke forthcoming.

52   Broadly speaking, by »complex discoveries«, I mean discoveries that require inter-, multi-, or transdisciplinary efforts. The discovery of anthropogenic climate change and the discovery of the coronavirus pandemic may be examples. I assume that further research on this topic will reveal that there is a wide range of complex, less complex, and uncomplex (simple) cases.

53   Phlogiston theory is a *locus classicus* when it comes to scientific revisability. According to phlogiston theory, which was accepted for more than 100 years, materials that burn contain a fire-like element which is released as the objects burn. Eventually, phlogiston theory was superseded by oxygen theory. Curiously, however, the discovery of oxygen took about 40 years, from 1772 to 1812 (see Wootton 2016, 86–87).

54   While the discovery of nuclear fission was clearly one of the most important scientific discoveries ever made, the discovery that Viagra reduces jet lag in hamsters on eastbound flights across time zones was arguably less important (Agostino, Plano & Golombek 2007).

55   As before, think of the discovery of oxygen which began in 1772 and ended in 1812 (Wootton 2016, 86–87). Another example may be the continental drift theory which Alfred Wegener first proposed as a hypothesis in lectures in 1912 and published in full in 1915; the theory was not accepted until the 1950s.

there is a lot of work to be done, and I want to highlight only some conceptual distinctions and subtleties that seem relevant here.

2.5.1 *Finding*

I believe that we cannot discover anything without also finding it. Therefore, I claim that finding is an indispensable feature of any discovery. This claim implies that finding is also an indispensable feature of any scientific discovery. Moreover, it seems that finding often marks the starting point of a scientific discovery process. Let us have a closer look.

In passing, I have already introduced a distinction with regard to the feature of finding which rests, technically speaking, on an understanding of finding as a binary relation – let us call it *sFo* – which says that the relatum s is the subject of finding (the finder) that finds the relatum o, which is the object of finding. The relation F that holds between these two relata is the relation of finding. This technical way of speaking helps us to see that we have three objects of investigation with regard to the feature of finding, namely the two relata and the relation that holds between them. I will discuss each of these briefly in turn.

As regards the first relatum, the subject of finding, we have already seen that this can either be a single individual, a team, or a group of individuals – joint discoveries are quite common in science. Since, beyond that, we live in an era of data-driven science, an era in which algorithms play an increasingly important role, we may also ask whether the subject of finding could also be a machine, maybe even in cooperation with humans (see Michel 2020b, Green & Michel forthcoming). At any rate, we have seen that the subject of finding may, but need not, exhibit further characteristics, such as having a eureka moment which can be brought about by observation or thought experimentation. It seems that having a eureka moment requires not only imaginative skillfulness and curiosity (Livio 2017), but also the ability to identify an object of finding *as such*. This, in turn, means that some background knowledge is needed on the part of the subject of finding (we will turn to the feature of knowledge below, § 2.5.3). Yet, the subject of finding does not always have to be very knowledgeable or even be an expert in the field in question. It can also happen that a layperson accidently comes across an object of finding that is later accepted as a new scientific discovery (we will turn to the feature of acceptance below as well, § 2.5.2). Depending on an individual discovery case, we can even distinguish between the finder and the describer of an object of finding.[56]

---

56   The finder-describer dichotomy is not unusual, e.g. in biology (Michel 2019, see also Ohl 2018). Thanks to Michael Ohl (Berlin) for helpful discussions.

As regards the object of finding, our second relatum, this can vary greatly. For one thing, the object of finding in a scientific discovery process can, ontologically speaking, be an individual, such as each of the astronomical objects that were discovered by members of the Herschel family (see § 2.2.1), most famously William Herschel's discovery of Uranus. The object of finding can, however, also consist of more than one individual, such as in the case of the amazing archaeological discovery of several Venetian glass beads that Mike Kunz and Robin Mills found in Alaska – possibly the first European objects to reach North America, predating Columbus' voyages (Kunz & Mills 2021). It can also be the case that the finding of one or more individuals marks the discovery of a new kind of physical entity (e.g., the Higgs particle, see § 2.4.4), a new biological species (e.g., *Lasiognathus dinema*, see Pietsch & Sutton 2015, Michel 2019), or even a so-called missing link (e.g., *Tiktaalik roseae*, see Shubin 2009). Moreover, in my view, the object of finding in a scientific discovery process could also be a fact (e.g., that the Earth revolves around the Sun), an interpretation (e.g., in literary studies), a method (e.g., in chemistry), or a proof (e.g., in mathematics). Of course, all of this needs to be explored in detail (see Münster 2022 for an attempt in this direction).

Finally, the relation of finding connects a subject of finding – or rather, its epistemic situation (belief system, knowledge) – with an object of finding. We can also say that it is a non-symmetrical relation, for an object of finding does generally not find its subject. There are, however, several open questions associated with the relation of finding as regards its proper characterization and its role in scientific discoveries. Among others, finding seems to be closely related to learning, which, in turn, is closely related to teaching – so, one open question associated with the relation of finding is this: What can we learn from analyzing the relation of finding with regard to science education?

2.5.2   *Acceptance*

As I have argued elsewhere (Michel 2019, Michel 2020b), finding is not alone sufficient for something to count as a scientific discovery – rather, it must also be accepted as a scientific discovery by a relevant scientific community. Imagine that, say, William Herschel never told anyone about what he observed with his telescope from the garden of his house in Bath, England, on 13 March 1781. It would certainly have been just that: a private person's observation, a simple case of finding, just as described above. Herschel's case would not have been much different from the case of Albert and the hole in his sock (see § 2.4.1 above). Fortunately, however, Herschel actually knew what he had to do and also did it: In April 1781, Herschel reported his observation to the Royal Society, the relevant scientific community in his case. If he had not done so, his finding would certainly not have gone down in history as the discovery of planet Uranus. The same holds, analogously, for other scientific discoveries. Now, my claim is this: There is no scientific discovery without acceptance. That is, I take acceptance, like finding, to be an indispensable or essential feature of any scientific discovery. But what does that mean?

First of all, the notion of acceptance generally requires that the people who are in the position(s) to decide whether something is to be accepted or not have access to the information in question. As regards scientific discoveries, this means that the relevant information concerning an object of finding is not to be kept private, but rather to be made publicly accessible to a relevant scientific community. Making something publicly accessible is literally the very idea of publication, which, in turn, has become an integral part of science over the centuries. One typical kind of scientific publication involves submitting a written piece of work (paper, book), which meets certain standards and criteria, about the finding to a relevant publishing outlet, such as a scientific journal.[57] In such cases, it is quite common that the piece of work is, initially, made accessible only to a small part of the scientific community which, depending on the scientific context and culture, may consist of colleagues, reviewers, or editors. Once the piece of work is – after a review procedure of whatever kind[58] – accepted for publication, it is made accessible to a greater part of the scientific community at least, and the finding in question counts as a new scientific discovery.

If we take into consideration (a) that it is an indispensable feature of science that we can make scientific discoveries, (b) that acceptance is an indispensable feature of scientific discoveries, and (c) that publication is a necessary precondition of acceptance, it is hardly surprising to find that we have established a historically grown, institutionalized, and diversified publication infrastructure in science. Clearly, the notion of publication is crucial to science and it serves different functions. Firstly, publication serves the function of quality control; for instance, we do not want every alleged finding immediately to count as a scientific discovery. Secondly, publication serves the function of informing the public, and, in doing so, it often also serves the function of recording and storing data. Thirdly, publication serves the function of enabling the appreciation of scientific achievements, such as a new scientific discovery.[59]

---

57   There are also other kinds of scientific publication (in the original sense of the term), such as oral presentations or panel discussions.

58   Depending on the context (e.g., field of discovery), review procedures can vary widely.

59   As many people working in science have learned the hard way, there is also huge room for improvement concerning different aspects of our scientific publication systems. To give but one example, think of publication bias (also known as the file drawer effect),

Against this background, the question arises of how we should characterize the transition from finding to acceptance within scientific discovery processes. I have already shown how a promising answer to this question could look like with the help of declarative speech acts and John Searle's account of institutional reality (Michel 2019). In a nutshell, I claim that language in general and declarative speech acts in particular play a crucial role within scientific discovery processes, leading to a picture of scientific discoveries as processes comprising non-institutional and institutional aspects. An important part of this picture is that the (individual) knowledge of a finding, once the finding is accepted as a new scientific discovery by a relevant part of a relevant scientific community, is integrated into the overall body of (institutional) knowledge of the discipline or field in question. Now, in order to get a better understanding of how we get from finding and acceptance to knowledge, let us, finally, turn to the third core feature of scientific discoveries.

2.5.3   *Knowledge*

As I see it, knowledge plays a double role within scientific discovery processes. For one thing, scientific discoveries can be seen as a path which leads us from finding and acceptance to knowledge. Seen this way, new scientific knowledge is the goal or end point of scientific discoveries. At the same time, we should not forget that we also need background knowledge in order to find anything in the first place. Seen this way, knowledge is the prerequisite or starting point of scientific discoveries. This already indicates that my picture of scientific discoveries is that of a dynamic process in which knowledge is both the starting point and the end point. This picture includes an accumulation view of scientific knowledge in the following sense: at the end of a scientific discovery process there is more knowledge than at the beginning – hence, scientific discoveries also contribute significantly to scientific progress.[60] Now, also because

---

which says that »studies that fail to show statistically significant effects, or that reproduce the work of others, have such low priority that they are effectively censored from the scientific record. They either end up in the file drawer or are never conducted in the first place« (Chambers 2019, 3). This is actually a big problem in a »culture where researchers must *publish* or *perish*« (ibid., 4, emphases in the original). Thanks to Mitch Green for discussion.

60   What I say here may initially seem to be very much in line with Bird's view of scientific progress (the epistemic account, see Bird 2007, Bird 2015). However, I also think that Dellsén has a point when he argues for a view of scientific progress based on understanding rather than knowledge (the noetic account, see Dellsén 2016, Dellsén 2018). Like the epistemic account, the noetic account also has its weaknesses, so that a combination of the epistemic and noetic accounts would be desirable. Goebel has made an interesting proposal for a hybrid account (Goebel 2019).

knowledge is a multifaceted key topic in epistemology as well as philosophy of science, I cannot go much into detail here. Still, I would like to highlight two distinctions that seem to be relevant for our purpose.

The first distinction I would like to highlight, following Gilbert Ryle's seminal work,[61] is not only a well-known but also fundamental one between two kinds of knowledge, namely knowledge-that and knowledge-how.[62] To illustrate, we are dealing with cases of knowledge-that when it is truly said of you that you know that some fact obtains, for instance, that the Earth revolves around the Sun, that there are eight planets in the Solar System, or that −40 °F is the same temperature as −40 °C. It is this kind of knowledge that has received by far the most attention in philosophy; it is even the classical answer to the question »What is knowledge?«. It is called »knowledge-that« because its contents can be expressed by means of »that«-clauses like in the examples above. According to the standard view (going back to Plato's *Theaetetus*), knowledge-that can be analyzed as justified true belief, which means that three conditions have to be fulfilled in order for something to be a case of knowledge.[63]

In contrast, we are dealing with cases of knowledge-how when it is truly said of you that you know how to do something, for instance, how to ride a bike, how to play the violin, or how to prepare a sample for viewing with a scanning electron microscope. It seems that what is known in these cases cannot simply be expressed by means of »that«-clauses and also that the standard analysis of

---

61   Ryle 1945 and Ryle 1949, chap. 2. See Stanley & Williamson 2001, but also Hornsby 2012.

62   As Fantl (2017) puts it, »the majority opinion in academic philosophy is that there is a considerable degree of independence« between these two kinds of knowledge. This majority opinion, however, has also been challenged since the mid-1970s. It seems that the first to argue that knowledge-how can be reduced to knowledge-that was Carl Ginet (1975). What is probably the most prominent elaboration of this line of argument has been brought forward by Jason Stanley and Timothy Williamson (Stanley & Williamson 2001, Stanley 2011). It should not go unnoticed, though, that the work of Stephen Boër and William Lycan (Boër & Lycan 1975, Boër & Lycan 1986) strongly influenced especially Stanley's work on the issue (see also Steup & Neta 2020, § 2.2). Over the last 20 years, several refinements as well as challenges to reductionist views have been brought forward. I cannot go into further detail concerning these debates here, but will, partly for the sake of convenience, follow the majority opinion mentioned above. (There are other reasons as well, see especially Kremer 2016.) Apart from that, whether you hold that knowledge-how is reducible to knowledge-that or not is, in my view, irrelevant to my picture of scientific discoveries. Thanks to Mitch Green and Bill Lycan for discussions and recollections.

63   As before, it is worth noting that the standard view has come under attack, most prominently by Edmund Gettier (1963). In addition, there are debates on issues related to the standard analysis of knowledge, such as epistemic luck (see Zagzebski 1994) or virtue-theoretic accounts (see Sosa 1980, Sosa 2007, Greco 2010). These debates, however, are not important for our purposes. Thanks to Ernest Sosa for earlier discussions of his views.

knowledge as justified true belief is not applicable here. So, knowledge-how appears to be not the same as knowledge-that but rather seems to consist in the reliable ability to engage well in an activity.[64]

The second distinction I would like to highlight is a fundamental one as well, even though it may not be as well-known. The general idea behind this distinction is present in several places in the literature, where we find, for instance, distinctions between personal and impersonal knowledge (Polanyi 1962), subjective and objective knowledge (Popper 1972), psychological and social knowledge (Hyman 1999), or natural and constitutive institutional knowledge (Tuomela 2011). We can say, broadly speaking, that all of these distinctions are based on the idea that we can think of knowledge in two fundamentally different ways.

For one thing, we can speak of knowledge in an individual sense. To illustrate, consider the following examples in which »K« is used to refer to an individual knower:

> (a) *K* knows that the cup is on the desk.
> (b) *K* knows how to brew a special beer.
> (c) *K* knows that the file is on the old computer.
> (d) *K* knows how to write a scientific paper on quantum mechanics.
> (e) *K* knows that the fish under the microscope is an exemplar of a newly discovered species.

Even though we probably first think of the individual knowers in these examples as human persons, I prefer to speak of individual knowledge rather than personal knowledge here. Among others, this *façon de parler* has the advantage that it leaves open the possibility that the knowers in question are neither persons nor human beings (e.g., animals, machines). I regard personal knowledge as a special kind of individual knowledge.

We can, however, also speak of knowledge in a non-individual sense. Here are a few examples:

> (f) It is common knowledge that Dublin is the capital of Ireland.
> (g) The literature study group in the English department knows how to interpret the works of Zora Neale Hurston.
> (h) We know today that smoking causes cancer.
> (i) NASA knows how to launch a rocket.
> (j) The world knows that anthropogenic climate change is real.

---

64   Wittgenstein comes to mind (1953, § 150): »The grammar of the word ›know‹ is evidently closely related to the grammar of the words ›can‹, ›is able to‹. But also closely related to that of the word ›understand‹. (To have ›mastered‹ a technique.)«.

In addition, we may sensibly ask »What did the university know about the harassment cases?« or »Does North Korea know how to build an atomic bomb?«. When it comes to the categorization and naming of all these different cases, there is, as far as I can see, no consensus in social epistemology – so, here is my preferred way of speaking: I regard all of the examples above as cases of social knowledge and some also as cases of institutional knowledge.[65] In other words: I regard institutional knowledge as a special kind of social knowledge.[66] What distinguishes institutional knowledge – such as, say, in the capital of Ireland, NASA, and university cases – from mere social knowledge is that institutional knowledge also requires acceptance (not just collective intentionality) as well as the existence of institutions.[67]

Now, by means of the two distinctions highlighted above, we arrive at the two-by-two matrix below. Since we are concerned with discoveries in science and since science, in the sense here understood, is institutionalized, we will focus on institutional knowledge.

	knowledge-that	knowledge-how
individual knowledge		
institutional knowledge		

As the matrix illustrates, there are four possible combinations. This is helpful to understand the different roles that knowledge plays in scientific discovery processes. First of all, as noted above, there is a difference between the starting point and the end point of scientific discoveries. Whereas we are dealing

---

65   Partly because much of classical epistemology has relied on an individualistic reading of the epistemic subject (see Goldman & O'Connor 2021, Bird 2010, 23–24), social epistemology is a comparatively young field of research (see Coady 1992, Schmitt 1994, and Goldman 1999; see also Nelson 1990 and Nelson 1993, thanks to Bill Lycan for pointing this out). Similar to the debates about knowledge-how and knowledge-that, one of the key questions in social epistemology debates is whether social knowledge is reducible to individual knowledge or not. In what follows, I assume that social knowledge is not the same as individual knowledge. My picture of scientific discoveries can be seen as a further development of Searle's account of social ontology (Searle 1995, Searle 2010), finding additional support in arguments from philosophy of science (such as in Bird 2014) and epistemology (such as in Dragos 2019).

66   This is analogous to Searle's distinction between social and institutional facts: »A special subclass of social facts are institutional facts, facts involving human institutions« (Searle 1995, 26).

67   Regarding the question of what an institution is, I am inclined to concur with Searle's response that an institution is a system of constitutive rules (Searle 2010, 10).

with individual scientists and their individual background knowledge (-that and -how) at the outset of scientific discovery processes, we end up with institutional knowledge (-that and -how) that each of us can access. Even though the following quote by John Hyman comes from a slightly different context, it nicely illustrates the point I want to make here:

> »We use the concept of knowledge to describe the cognitive condition of individuals; but we also use it to describe the progress of scientific and historical research. So for example we can speak or enquire about the state of knowledge in a particular field of biology or history. And if we do so, we are evidently not concerned with what anyone in particular knows about, say, the genetics of fruit flies or the career of Charlemagne, but rather with what the scientific or academic community knows. Needless to say, there is a close connection between [individual] and [institutional] knowledge. But ›It is known that p‹ does not simply mean ›Someone or other knows that p‹. To suppose that it does is to ignore the role of documents, archives and libraries in the economy of knowledge.« (Hyman 1999, 433–434)

Moreover, the distinction between knowledge-that and knowledge-how proves useful in so far as it allows us to distinguish between different objects of discovery. It seems that discoveries of, say, new individuals, species, and facts lead to new knowledge-that, whereas discoveries of new methods lead to new knowledge-how. For instance, William Herschel's discovery of Uranus led to new knowledge-that – we now know that this celestial body is a planet –, whereas the discovery of a method of diagnosing a broad range of cancers at their early stages by utilizing a particular malaria protein that sticks to cancer in blood samples made by researchers from Copenhagen led to new knowledge-how – we now know how to detect certain types of cancer at an early stage (Agerbæk, Bang-Christensen, Yang, et al. 2018).

Concluding Remarks

In this essay, I have claimed that we need a philosophy of scientific discovery. Before turning to the question of what such a philosophy might look like, however, we addressed two basic questions: Don't we have a philosophy of scientific discovery yet? And do we need one at all? To answer the first question, we took a closer look at history and found that we have not had a systematic philosophy of scientific discovery worthy of the name for over 150 years (§ 2.2). To answer the second question, I put forward three arguments that show the importance of a philosophy of scientific discovery (§ 2.3). In a nutshell, we arrived at the following answers: No, we do not yet have a philosophy of scientific discovery, and yes, we definitely need one.

It is time to remedy this shortcoming. To this end, we first took a closer look at the concept of discovery (§ 2.4), leading us to the insight that scientific discoveries have an underlying structure with certain structural features. We found that there are several features of scientific discoveries that may be important or even distinctive of, say, a certain field of research, but that are not indispensable or essential to scientific discovery processes as such. These features include, among others, eureka moments, serendipities, joint discoveries, special science funding, or the use of certain instruments (e.g., microscopes and telescopes). In addition, we found three indispensable or essential structural features which we examined in more detail (§ 2.5) and which we placed in a picture with a certain dynamics: the process of making scientific discoveries can be seen as a path, leading us from finding and acceptance to knowledge.

How should we proceed in the philosophy of scientific discovery? On the one hand, it is certainly worthwhile to elaborate on the approach that I outlined in more detail in order to gain a better understanding of scientific discovery processes in a wider context: How do the individual structural features of scientific discovery processes relate to each other? What roles do language and institutions play? What can we learn from all this in terms of improving our scientific publication, education, and funding systems? On the other hand, we should think about how to approach the philosophy of scientific discovery itself in a structured way. For starters, it would be helpful to distinguish not only between a historical and a systematic dimension of the philosophy of scientific discovery, but also between general and special philosophy of scientific discovery – analogous to the distinction between general and special philosophy of science.

I regard this essay as a first step toward a philosophy of scientific discovery – which is itself still uncharted territory, with much yet to be discovered and explored.

Acknowledgments

Thanks to Hanne Andersen, Paavo Bergmann, Kim Boström, Lukas Brand, Jörg Friedrich, Benedikt Göcke, Mitch Green, Susanne Hahn, Michael Kienecker, Florian Lippert, Mario Livio, Bill Lycan, Joe Moore, Gernot Münster, Michael Ohl, Christian Pelz, Michael Pohl, Stefan Reins, Theo Michael Schmitt, Niko

Strobach, Christian Tapp, Holm Tetens, Gottfried Vosgerau, and several audiences for helpful discussion. Special thanks to The Herschel Family Archive, The Herschel Museum of Astronomy, and especially to Will Herschel-Shorland for supporting my work. Research for this article was funded by the German Research Foundation (Deutsche Forschungsgemeinschaft, project number 295845819).

References

Agerbæk, M.Ø., Bang-Christensen, S.R., Yang, MH., et al. 2018. »The VAR2CSA Malaria Protein Efficiently Retrieves Circulating Tumor Cells in an EpCAM-Independent Manner.« *Nature Communications* 9: 3279.

Agostino, P.V., S.A. Plano, & Golombek, D.A. 2007. »Sildenafil Accelerates Reentrainment of Circadian Rhythms after Advancing Light Schedules.« *Proceedings of the National Academy of Sciences* 104 (23): 9834–9839.

Appleby, J. 2013. *Shores of Knowledge: New World Discoveries and the Scientific Imagination.* New York: Norton.

Barlow, N., ed. 1958. *The Autobiography of Charles Darwin, 1809–1882.* New York: Norton.

Barlow, N., ed. 1967. *Darwin and Henslow: The Growth of an Idea. Letters 1831–1860,* Berkeley: University of California Press.

Bird, A. 2007. »What Is Scientific Progress?« *Noûs* 41: 92–117.

Bird, A. 2010. »Social Knowing: The Social Sense of ›Scientific Knowledge.‹« *Philosophical Perspectives* 24 (1): 23–56.

Bird, A. 2014. »When Is There a Group that Knows? Distributed Cognition, Scientific Knowledge, and the Social Epistemic Subject.« In *Essays in Collective Epistemology,* edited by J. Lackey, 42–63. Oxford: Oxford University Press.

Bird, A. 2015. »Scientific Progress.« In *The Oxford Handbook of Philosophy of Science,* edited by P. Humphreys, 544–563. Oxford: Oxford University Press.

Blackburn, S. 2016. *The Oxford Dictionary of Philosophy.* 3rd ed. Oxford: Oxford University Press.

Boër, S.E. & Lycan, W.G. 1975. »Knowing Who.« *Philosophical Studies* 28 (5): 299–344.

Boër, S.E. & Lycan, W.G. 1986. *Knowing Who.* Cambridge, Mass.: MIT Press.

Brown, J.R. & Fehige, Y. 2019. »Thought Experiments.« In *The Stanford Encyclopedia of Philosophy,* edited by E.N. Zalta. plato.stanford.edu/archives/win2019/entries/thought-experiment/.

Brämer, U. 2020 »Spektakuläre Riesenkristallfunde im Windloch.« https://akkh.de/spektakulaere-riesenkristallfunde-im-windloch/.

Buttmann, G. 1965. *John Herschel: Lebensbild eines Naturforschers*. Stuttgart: Wissenschaftliche Verlagsgesellschaft.

Cannon, W.F. 1961. »John Herschel and the Idea of Science.« *Journal of the History of Ideas* 22 (2): 215–239.

Cannon, W.F. 1967. »John Herschel.« In *Encyclopedia of Philosophy*, vol. 3, edited by P. Edwards, New York: Collier & Macmillan.

Cannon, W.F. 1970a. »Giving a Man his Due.« *Science* 168 (3932): 731–732.

Cannon, W.F. 1970b. »Mistaken Identity.« *Science* 169 (3941): 128.

Chambers, C. 2019. *The Seven Deadly Sins of Psychology: A Manifesto for Reforming the Culture of Scientific Practice*. Princeton & Oxford: Princeton University Press.

Coady, C.A.J. 1992. *Testimony: A Philosophical Study*. Oxford: Clarendon Press.

Copeland, S.M. 2017. »On Serendipity in Science: Discovery at the Intersection of Chance and Wisdom.« *Synthese* 196: 2385–2406.

Copeland, S.M. 2018. »›Fleming Leapt on the Unusual like a Weasel on a Vole‹: Challenging the Paradigms of Discovery in Science.« *Perspectives on Science* 26 (6): 694–721.

Cudnik, B. 2013. *Faint Objects and How to Observe Them*. Dordrecht: Springer.

Darwin, C. 1859. *On the Origin of Species*. In *Darwin Online*, edited by J. v. Wyhe. http:// darwin-online.org.uk/.

Dellsén, F. 2016. »Scientific Progress: Knowledge versus Understanding.« *Studies in History and Philosophy of Science* 56: 72–83.

Dellsén, F. 2018. »Scientific Progress: Four Accounts.« *Philosophy Compass* 13 (11): 1–10.

de Rond, M. & Morley, I., eds. 2010. *Serendipity: Fortune and the Prepared Mind*. Cambridge: Cambridge University Press.

Dispenza, F., F. Cappello, G. Kulamarva, & A. De Stefano. 2013. »The Discovery of Stapes.« *Acta Otorhinolaryngologica Italica* 33 (5): 357–359.

Dragos, C. 2019. »Groups Can Know How.« *American Philosophical Quarterly* 56 (3): 265–276.

Duhem, P. [1908] 1998. *Ziel und Struktur der physikalischen Theorien*. Hamburg: Meiner.

Ebenau, C. & Voigt, S. 2019. »Große Neuentdeckung im Bergischen Land (NRW).« *Mitteilungen des Verbandes der deutschen Höhlen- und Karstforscher* 65 (1 & 2): 39.

Eco, U. 1999. *Serendipities: Language and Lunacy*. London: Phoenix.

Einstein, A. 1905. »Zur Elektrodynamik bewegter Körper.« *Annalen der Physik* 322 (10): 891–921.

Einstein, A. 1949. »Autobiographical Notes.« In *Albert Einstein: Philosopher-Scientist*, edited by P.A. Schilpp, 1–95. Evanston: Open Court.

Einstein, A. [1917] 2009. *Über die spezielle und die allgemeine Relativitätstheorie*. 24th ed. Berlin & Heidelberg: Springer.

Fantl, J. 2017. »Knowledge How.« In *The Stanford Encyclopedia of Philosophy* (Fall 2017 Edition), edited by E.N. Zalta. https://plato.stanford.edu/archives/fall2017/entries/knowledge-how/.

Field, H.W. 1871. »Obituary Notice of Sir John Frederick William Herschel.« *Proceedings of the American Philosophical Society* 12 (86): 217–223.

Frege, G. [1879–1891] 1983. »Logik.« In *Gottlob Frege: Nachgelassene Schriften*, edited by H. Hermes, F. Kambartel, & F. Kaulbach, 1–8. Hamburg: Meiner.

Gettier, E.L. 1963. »Is Justified True Belief Knowledge?« *Analysis* 23 (6): 121–123.

Gibson, S. 2019. *The Spirit of Inquiry: How One Extraordinary Society Shaped Modern Science*. Oxford: Oxford University Press.

Ginet, C. 1975. *Knowledge, Perception and Memory*. Dordrecht: Springer.

Goebel, C. 2019. »A Hybrid Account of Scientific Progress: Finding Middle Ground Between the Epistemic and the Noetic Accounts.« *Kriterion – Journal of Philosophy* 33 (3): 1–16.

Goldman, A. 1999. *Knowledge in a Social World*. Oxford: Oxford University Press.

Goldman, A. & O'Connor, C. 2021. »Social Epistemology.« In *The Stanford Encyclopedia of Philosophy* (Spring 2021 Edition), edited by E.N. Zalta. https://plato.stanford.edu/archives/spr2021/entries/epistemology-social/.

Greco, J. 2010. *Achieving Knowledge: A Virtue-Theoretic Account of Epistemic Normativity*. Cambridge: Cambridge University Press.

Greene, B. 2013. »How the Higgs Boson Was Found.« *Smithsonian Magazine*. July 2013. https://www.smithsonianmag.com/science-nature/how-the-higgs-boson-was-found-4723520/.

Green, M.S. & Michel, J.G. (forthcoming). »What Might Machines Mean?« *Minds and Machines*.

Haack, S. [1980] 2019. »Personal or Impersonal Knowledge?« *Journal of Philosophical Investigations* 13 (28): 20–44.

Halacy, D. 1970. *Charles Babbage: Father of the Computer*. New York: Crowell-Collier Press.

Hanson, N.R. 1958a. *Patterns of Discovery: An Inquiry into the Conceptual Foundations of Science*. Cambridge: Cambridge University Press.

Hanson, N.R. 1958b. »The Logic of Discovery.« *The Journal of Philosophy* 55 (25): 1073–1089.

Hanson, N.R. 1967. »An Anatomy of Discovery.« *The Journal of Philosophy* 64 (11): 321–352.

Hanson, N.R. 1969, *Perception and Discovery: An Introduction to Scientific Inquiry*, edited by W.C. Humphreys. San Francisco: Freeman, Cooper & Co.

Hartshorne, C. 1965. *Anselm's Discovery: A Re-Examination of the Ontological Proof of God's Existence*. LaSalle, Ill.: Open Court.

Hausman, D.M. 2003. »Book Review: Philip Kitcher, *Science, Truth, and Democracy*.« *Ethics* 113 (2): 423–428.

Heis, J. 2018. »Neo-Kantianism.« In *The Stanford Encyclopedia of Philosophy* (Summer 2018 Edition), edited by E.N. Zalta. https://plato.stanford.edu/archives/sum2018/entries/neo-kantianism/.

Herschel, J.F.W. 1831. *A Preliminary Discourse on the Study of Natural Philosophy*. https://www.biodiversitylibrary.org/page/17907244.

Herschel, J.F.W. 1868. »The Reverend William Whewell, DD.« *Proceedings of the Royal Society of London* 65: li–lxi.

Hilbert, D. 1900. »Mathematische Probleme. Vortrag, gehalten auf dem internationalen Mathematiker-Kongreß zu Paris 1900.« In *Nachrichten von der Gesellschaft der Wissenschaften zu Göttingen, Mathematisch-Physikalische Klasse*, 253–297. http://www.digizeitschriften.de/dms/img/?PID=GDZPPN002498863.

Honenberger, P. 2018. »Darwin among the Philosophers: Hull and Ruse on Darwin, Herschel, and Whewell.« *Hopos: The Journal of the International Society for the History of Philosophy of Science* 8 (2): 278–309.

Hornsby, J. 2012. »Ryle's Knowing-How, and Knowing How to Act.« In *Knowing How: Essays on Knowledge, Mind, and Action*, edited by J. Bengson & M.A. Moffett, 80–100. Oxford: Oxford University Press.

Hoskin, M. 2005. »Herschel, Caroline Lucretia (1750–1848).« In *Oxford Dictionary of National Biography*. https://doi.org/10.1093/ref:odnb/13100.

Hurley, M.M., D.C. Dennett, & R.B. Adams, Jr. 2011. *Inside Jokes: Using Humor to Reverse-Engineer the Mind*. Cambridge, Mass.: MIT Press.

Hyman, J. 1999. »How Knowledge Works.« *The Philosophical Quarterly* 49 (197): 433–451.

Kantorovich, A. 1993. *Scientific Discovery: Logic and Tinkering*. Albany: State University of New York Press.

Kitcher, P. 2001. *Science, Truth, and Democracy*. Oxford: Oxford University Press.

Kremer, M. 2016. »A Capacity to Get Things Right: Gilbert Ryle on Knowledge.« *European Journal of Philosophy* 25 (1): 25–46.

Kunz, M.L., & Mills, R.O. 2021. »A Precolumbian Presence of Venetian Glass Trade Beads in Arctic Alaska.« *American Antiquity*. Published online 20 January 2021, doi:10.1017/aaq.2020.100.

Liu, A. 2012. »The Humanities and Tomorrow's Discoveries.« *4Humanities*, 25 July 2012. https://4humanities.org/2012/07/alan-liu-the-humanities-and-tomorrows-discoveries/.

Livio, M. 2013. *Brilliant Blunders*. New York: Simon & Schuster.

Livio, M. 2017. *Why? What Makes Us Curious*. New York: Simon & Schuster.

Livio, M. 2020. *Galileo and the Science Deniers*. New York: Simon & Schuster.

Livio, M. 2022. »On the Role of Brilliant Blunders in Science.« This volume.

Losee, J. 2001. *A Historical Introduction to the Philosophy of Science*. 4th ed. Oxford: Oxford University Press.

Lund, M. 2010. *N.R. Hanson: Observation, Discovery, and Scientific Change*. Amherst, NY: Humanity Books.

Mach, E. 1896. »On the Part Played by Accident in Invention and Discovery.« *The Monist* 6 (2): 161–175.

Mach, E. [1905] 1917. *Erkenntnis und Irrtum: Skizzen zur Psychologie der Forschung*. Leipzig: Johann Ambrosius Barth.

Macleod, C. 2020. »John Stuart Mill.« In *The Stanford Encyclopedia of Philosophy* (Summer 2020 Edition), edited by E.N. Zalta. https://plato.stanford.edu/archives/ sum2020/entries/mill/.

Meheus, J. 1999. »The Positivists' Approach to Scientific Discovery.« *Philosophica* 64 (2): 81–108.

Meheus, J., & Nickles, T., eds. 2009. *Models of Discovery and Creativity*. Dordrecht: Springer.

Merton, R.K., & Barber, E. [1958] 2004. *The Travels and Adventures of Serendipity*. Princeton: Princeton University Press.

Michel, J.G. 2011. *Der qualitative Charakter bewusster Erlebnisse: Physikalismus und phänomenale Eigenschaften in der analytischen Philosophie des Geistes*. Münster: mentis.

Michel, J.G. 2013. »Mit Gedankenexperimenten argumentieren: Eine Fallstudie in der Philosophie des Geistes.« In *Die Suche nach dem Geist*, edited by J.G. Michel & G. Münster, 81–120. Münster: mentis.

Michel, J.G. 2019. »How Are Species Discovered? Declarative Speech Acts in Biology.« *Grazer Philosophische Studien* 96 (3): 419–441.

Michel, J.G. 2020a. »Imaginary Demons and Scientific Discoveries.« *Science* 370 (6518): 772.

Michel, J.G. 2020b. »Could Machines Replace Human Scientists? Digitalization and Scientific Discoveries.« In *Artificial Intelligence: Reflections in Philosophy, Theology, and the Social Sciences*, edited by B.P. Göcke & A. Rosenthal-von der Pütten, 361–376. Paderborn: mentis/Brill.

Michel, J.G., & Göcke, B.P. (forthcoming). »Scientific Discoveries in Theology?«

Mill, J.S. [1843] 1974. *A System of Logic Ratiocinative and Inductive: Being a Connected View of the Principles of Evidence and the Methods of Scientific Investigation*. In *The Collected Works of John Stuart Mill*, vol. VIII, edited by J.M. Robson. Toronto: University of Toronto Press. http://oll-resources.s3.amazonaws.com/ titles/247/0223.08_Bk.pdf.

Mudry, A. 2013. »Disputes Surrounding the Discovery of the Stapes in the Mid 16th Century.« *Otology & Neurotology* 34 (3): 588–592.

Münster, G. 2022. "Discovering, Inventing, Contriving – What Are Scientists Actually Doing?" This volume.

Naraniecki, A. 2010. »Neo-Positivist or Neo-Kantian? Karl Popper and the Vienna Circle.« *Philosophy* 85 (4): 511–530.

Nelson, L.H. 1990. *Who Knows? From Quine to a Feminist Empiricism.* Philadelphia: Temple University Press.

Nelson, L.H. 1993. »Epistemological Communities.« In *Feminist Epistemologies,* edited by L. Alcoff & E. Potter, 121–159. New York: Routledge.

Nickles, T., ed. 1980a. *Scientific Discovery: Case Studies. Boston Studies in the Philosophy of Science,* vol. 60. Dordrecht: Springer.

Nickles, T., ed. 1980b. *Scientific Discovery, Logic, and Rationality, Boston Studies in the Philosophy of Science,* vol. 56. Dordrecht: Springer.

Nickles, T. 1980c. »Introductory Essay: Scientific Discovery and the Future of Philosophy of Science.« In *Scientific Discovery, Logic, and Rationality,* edited by T. Nickles, *Boston Studies in the Philosophy of Science,* vol. 56, 1–59, Dordrecht: Springer.

Ohl, M. 2018. *The Art of Naming.* Translated by E. Lauffer. Cambridge, Mass.: MIT Press.

Pence, C.H. 2018. »Sir John F.W. Herschel and Charles Darwin: Nineteenth-Century Science and Its Methodology.« *HOPOS* 8 (1), 108–140.

Pietsch, T.W. & Sutton, T.T. 2015. »A New Species of the Ceratioid Anglerfish Genus *Lasiognathus* Regan (Lophiiformes: Oneirodidae) from the Northern Gulf of Mexico.« *Copeia* 103 (2): 429–432.

Polanyi, M. 1962. *Personal Knowledge: Towards a Post-Critical Philosophy.* Corr. ed. London: Routledge.

Popper, K.R. 1934. *Logik der Forschung. Zur Erkenntnistheorie der modernen Naturwissenschaften.* Wien: Springer. Online source: https://monoskop.org/images/8/83/Popper_Karl_Logik_der_Forschung_1935.pdf. Translated into English as *The Logic of Scientific Discovery.* London, 1959.

Popper, K.R. 1935. »›Induktionslogik‹ und ›Hypothesenwahrscheinlichkeit.‹« *Erkenntnis* 5: 170–172.

Popper, K.R. 1972. *Objective Knowledge: An Evolutionary Approach.* Oxford: Clarendon Press.

Reichenbach, H. [1938] 2006. *Experience and Prediction: An Analysis of the Foundations and the Structure of Knowledge.* Notre Dame: University of Notre Dame Press.

Ross, S. 1962. »*Scientist:* The Story of a Word.« *Annals of Science* 18: 65–85.

Ruse, M. 1975. »Darwin's Debt to Philosophy: An Examination of the Influence of the Philosophical Ideas of John F. W. Herschel and William Whewell on the Development of Charles Darwin's Theory of Evolution.« *Studies in History and Philosophy of Science* 6 (2): 159–181.

Ruse, M. 1978. »Discussion: Darwin and Herschel.« *Studies in History and Philosophy of Science* 9 (4): 323–331.

Ruse, M. 2010. »David Hull: A Memoir.« *Biology & Philosophy* 25: 739–747.

Ryle, G. 1945. »Knowing How and Knowing That.« *Proceedings of the Aristotelian Society* 46: 1–16.

Ryle, G. 1949. *The Concept of Mind*. Chicago: University of Chicago Press.

Sarkar, S. & Pfeifer, J., eds. 2006. *The Philosophy of Science: An Encyclopedia*. New York: Routledge.

Schmitt, F.F. 1994. *Socializing Epistemology: The Social Dimensions of Knowledge*. Lanham, Md.: Rowman & Littlefield.

Schrenk, M. 2004. »Galileo versus Aristotle on Free Falling Bodies.« *History of Philosophy & Logical Analysis* 7 (1): 81–89.

Schweber, S.S. 1989. »John Herschel and Charles Darwin: A Study in Parallel Lives.« *Journal of the History of Biology* 22 (1): 1–71.

Searle, J.R. 1995. *The Construction of Social Reality*. London: Penguin.

Searle, J.R. 2010. *Making the Social World: The Structure of Human Civilization*. Oxford: Oxford University Press.

Shorland, E. 2016. *The Forgotten Philosopher: Sir John F.W. Herschel*, edited by W. Herschel-Shorland. Published by The Herschel Family Archive.

Shubin, N. 2009. *Your Inner Fish*. London: Penguin.

Simenhoff, J. 1963. »Sir William and Sir John Herschel.« *Monthly Notes of the Astronomical Society of Southern Africa* 22: 114–126. http://articles.adsabs.harvard.edu/pdf/1963MNSSA..22..114S.

Snyder, L.J. 1997. »Discoverer's Induction.« *Philosophy of Science* 64 (4): 580–604.

Snyder, L.J. 2006. *Reforming Philosophy: A Victorian Debate on Science and Society*. Chicago: University of Chicago Press.

Snyder, L.J. 2011. *The Philosophical Breakfast Club: Four Remarkable Friends Who Transformed Science and Changed the World*. New York: Broadway Books.

Snyder, L.J. 2019. »William Whewell.« In *The Stanford Encyclopedia of Philosophy* (Spring 2019 Edition), edited by E.N. Zalta. https://plato.stanford.edu/archives/spr2019/entries/whewell/.

Sosa, E. 1980. »The Raft and the Pyramid: Coherence versus Foundations in the Theory of Knowledge.« *Midwest Studies in Philosophy* 5 (1): 3–25.

Sosa, E. 2007. *A Virtue Epistemology: Apt Belief and Reflective Knowledge*, vol. 1. Oxford: Oxford University Press.

Standage, T. 2000. *The Neptune File: A Story of Astronomical Rivalry and the Pioneers of Planet Hunting*. New York: Walker & Company.

Stanley, J. & Williamson, T. 2001. »Knowing How.« *Journal of Philosophy* 98 (8): 411–444.

Stanley, J. 2011. *Know How*. Oxford: Oxford University Press.

Steup, M. & Neta, R. 2020. »Epistemology.« In *The Stanford Encyclopedia of Philosophy* (Fall 2020 Edition), edited by E.N. Zalta. https://plato.stanford.edu/archives/fall2020/entries/epistemology/.

Thagard, P.R. 1977. »Discussion: Darwin and Whewell.« *Studies in History and Philosophy of Science* 8 (4): 353–356.

Todhunter, I., ed. [1876] 2011. *William Whewell, Master of Trinity College, Cambridge: An Account of his Writings*, vol. 2. Cambridge: Cambridge University Press.

Tuomela, R. 2003. »Collective Acceptance, Social Institutions, and Social Reality.« *American Journal of Economics and Sociology* 62 (1): 123–165.

Tuomela, R. 2004. »Group Knowledge Analyzed.« *Episteme* 1 (2): 109–127.

Tuomela, R. 2011. »An Account of Group Knowledge.« In *Collective Epistemology*, edited by H.B. Schmid, D. Sirtes, & M. Weber, 75–117. Heusenstamm: Ontos/De Gruyter.

Warner, B. 2009. »Charles Darwin and John Herschel.« *South African Journal of Science* 10: 432–439.

Warner, B. & Rourke, J. 1996. *Flora Herscheliana: Sir John and Lady Herschel at the Cape 1834 to 1838.* Johannesburg: Brenthurst Press.

Weinberg, S. 2015. *To Explain the World: The Discovery of Modern Science.* London: Penguin.

Whewell, W. 1831. »Review of J. Herschel's *Preliminary Discourse on the Study of Natural Philosophy.*« *Quarterly Review* 90: 374–407.

Whewell, W. [1840] 1996. *The Philosophy of the Inductive Sciences*, vol. II. London: Routledge/Thoemmes.

Whewell, W. 1847. *The Philosophy of the Inductive Sciences, Founded Upon Their History.* 2nd edition, in two volumes. London: John W. Parker.

Whewell, W. 1849. *Of Induction, With Especial Reference to Mr. J. Stuart Mill's System of Logic.* London: John W. Parker.

Whewell, W. 1858a. *The History of Scientific Ideas.* London: John W. Parker.

Whewell, W. 1858b. *Novum Organon Renovatum.* London: John W. Parker.

Whewell, W. 1860. *On the Philosophy of Discovery: Chapters Historical and Critical.* London: John W. Parker.

Whitrow, G.J. 1989. »Newton's Role in the History of Mathematics.« *Notes and Records of the Royal Society of London* 43 (1): 71–92.

Wittgenstein, L. 1953. *Philosophische Untersuchungen. Philosophical Investigations.* The German text, with an English translation by G.E M. Anscombe, P.M.S. Hacker, & J. Schulte. Revised 4th ed. Malden, Mass.: Wiley-Blackwell, 2009.

Wootton, D. 2016. *The Invention of Science: A New History of the Scientific Revolution.* London: Penguin.

Wootton, D. 2017. »History: Science and the Reformation.« *Nature* 550: 454–455.

Yaqub, O. 2018. »Serendipity: Towards a Taxonomy and a Theory.« *Research Policy* 47 (1): 169–179.

Zagzebski, L. 1994. »The Inescapability of Gettier Problems.« *The Philosophical Quarterly* 44 (174): 65–73.

On the Role of Brilliant Blunders in Science

Mario Livio

3.1   Introduction

In 1991, astronomers Andrew Lyne, Matthew Bailes and S. L. Shemar made what was regarded then as an electrifying announcement: the discovery of the very first planet outside our Solar System (Bailes, Lyne, & Shemar 1991).

To everyone's surprise, however, this planet was not found orbiting another Sun-like star, but rather around a pulsar – the extremely dense and rapidly spinning neutron-star remnant of a supernova explosion. Stars more massive than about eight solar masses end their lives in a powerful explosion that ejects the star's outer layers at speeds of tens of thousands miles per second, leaving behind their collapsed core – a neutron star.

The putative planet was claimed to have been detected by the very slight changes it introduced into the period of the radio-frequency pulses emitted by the spinning pulsar. Since those pulses are measured with an extraordinarily high precision, even tiny deviations from perfect periodicity can, in principle, be observed and interpreted.

Unexpectedly, Lyne and Bailes had to retract the discovery a few months later, after uncovering an error in their analysis (Lyne & Bailes 1992). The astronomers admitted that they had not corrected adequately for the Earth's motion around the Sun in their measurements. Lyne presented the error in detail at a meeting of the American Astronomical Society in January 1992, and the public acknowledgement of the blunder won him a standing ovation from the astronomical community. Fortunately for the search for extrasolar planets, in this case the tale had a surprising ending.

Immediately following Lyne's presentation, astronomer Aleksander Wolszczan announced that he and his colleague Dale Frail had truly discovered two planets orbiting another pulsar, using the same technique. These turned out to be indeed the very first known extrasolar planets. Wolszczan told me later that Lyne's original paper had acted as a »confidence booster«, which convinced him that the evidence for planets in his own data was real. By the time Lyne withdrew his result, Wolszczan had already performed a sufficient number of checks and tests to be absolutely certain.

Most researchers know that blunders are, in some sense, an integral part of the scientific process. Research is not a linear march to the truth but rather a

© BRILL MENTIS, 2022 | DOI:10.30965/9783957437044_004

zigzag path, involving many false starts and blind alleys. Furthermore, mistakes are definitely not the exclusive province of sloppy, not sufficiently thoughtful, or inexperienced scientists. As I have shown in my book *Brilliant Blunders* (Livio 2013), even the brightest luminaries – including Charles Darwin and Albert Einstein – made some serious blunders.

Truly innovative ideas require a willingness to embrace risks and an acceptance of the fact that errors can be portals to progress. Although this is well known to many private companies engaged in research and development (in particular to most start-up endeavors), academic institutions are still somewhat slow in recognizing the almost necessity of blunders. Still, the famous chemist and Nobel Laureate Linus Pauling knew it well. His former postdoc, chemist and crystallographer Jack Dunitz, recalled being told by Pauling: »Mistakes do no harm in science because there are lots of smart people out there who will immediately spot a mistake and correct it. You can only make a fool of yourself and that does no harm, except to your pride. If it happens to be a good idea, however, and you don't publish it, science may suffer a loss.« Pauling himself followed his own advice. He published his wrong triple-helix model of DNA. Watson and Crick, who used precisely Pauling's methodology, were able to spot his mistake and come up with the correct double-helix model.

3.2 To Be or Knot to Be

Even preposterous ideas can sometimes lead to important insights. In 1867, the eminent physicist William Thomson (Lord Kelvin) proposed that atoms were not point-like entities but ›knotted vortex tubes of the ether‹ (Thomson 1867). Ether was the fluid that was supposed to pervade all space, providing a medium for electricity and magnetism to propagate in. Inspired by work on vortices in fluids by the nineteenth-century German physicist Hermann von Helmholtz, Kelvin identified three characteristics of knotted vortex tubes that, in his view, made them attractive models for atoms.

First, vortices in fluids were astonishingly stable, mirroring to Kelvin the »unalterable distinguishing qualities« of atoms. Second, Kelvin thought that the variety of chemical elements could reflect the »endless variety« of knots. Finally, Kelvin surmised, just as smoke rings could be made to vibrate, the oscillations of ether vortex tubes might produce the observed atomic spectral lines.

To be able to really explain the periodic table, however, Kelvin needed a way to classify knots according to their forms (e.g., the number of crossings they involved when projected onto a plane), discarding any that could be manipulated from one form to another. In Kelvin's theory, the circular ›unknot‹ (without any crossing) represented the hydrogen atom; the triple-looped ›trefoil‹ knot, the carbon atom; the figure-eight knot, the oxygen atom, and so on.

Mathematician Peter Guthrie Tate took upon himself the Herculean task of classifying knots. Mathematical knots look precisely like familiar knots in a string, only with the string's ends spliced. That is, a mathematical knot is a closed curve with no loose ends. After laboring for about two decades Tate was able to publish tables of knots with only up to ten crossings. By then, however, it had become clear that Kelvin's theory of vortex atoms was wrong. Moreover, physicists Albert Michelson and Edward Morley demonstrated later that the ether does not even exist. But this failure of Kelvin's theory did not deter everyone from being fascinated by knots. Whereas physicists indeed lost interest for a while, knots began to intrigue mathematicians, becoming an active area of study for decades.

Unexpectedly, in the 1980s, knot theory reconnected back with physics. Mathematician Vaughan Jones discovered a polynomial – an algebraic expression – that is unique for every knot. Mathematical physicist Edward Witten linked the so-called *Jones Polynomial* to quantum field theory, the branch of physics that describes fields and the subatomic world. In classical physics, the path of a particle traveling from point A to point B is uniquely determined by Newton's laws of motion. In the quantum regime, on the other hand, one has to consider all the possible paths connecting A to B, including winding and knotted ways.

Subsequent work further linked knots, quantum field theory and *string theory*, which, by describing particles as vibrations of entities called strings, harks in some sense back to Kelvin's idea. Today, knots are also used in chemistry and biology to analyze the actions of enzymes on DNA molecules. In a process known as site-specific recombination, enzymes align segments of the genetic sequence, cut the two strands of DNA open and recombine the four ends in various ways – a process which can be mathematically described using knot theory.

3.3   Different Molecules of Life?

Blunders are sometimes hard to correct. Modern scientific experiments can be so intricate and may require such big investments in both time and funds that replicating them becomes almost prohibitive. Consequently, when a result is widely believed to be wrong, few scientists are motivated to repeat the work.

But sometimes there can be rewards for bothering to do so. The sensational claim in the journal *Science* by geomicrobiologist Felisa Wolfe-Simon and her colleagues, that they have discovered a bacterium that substitutes arsenic for phosphorus in its DNA to sustain its growth, brought about a wave of criticism.

A few of the critics even took on the onerous task of carefully checking the experiment. In particular, microbiologist Rosemary Redfield at the University of British Columbia in Vancouver, Canada, not only repeated the study but also blogged in detail about the process (Redfield 2012). The effort proved fruitful, showing that the bacterium goes to great lengths to actually dodge arsenic. Redfield and her colleagues detected no arsenic in the bacterium's DNA, measuring it down to much lower levels than the claimed detections in the original paper. Molecular biologist Dan Tawfik and his team at the Weizmann Institute of Science in Rehovot, Israel, identified the mechanism by which some of the proteins of this and related bacteria bind to phosphate and not to arsenate (Tawfik & Viola 2011).

Although one lesson is obvious – extraordinary claims require extraordinary evidence – the original paper still had some scientific value. It stimulated discussion and inspired curiosity about different possibilities for life forms.

3.4   Brilliant Blunders

In the nineteenth century, Scottish author Samuel Smiles wrote: »We often discover what will do, by finding out what will not do; and probably he who never made a mistake never made a discovery.« His statement should not be taken as advocacy for slapdash science but as an encouragement to think along original paths, and occasionally take calculated risks.

Can research failure be accommodated in today's fast-paced, funding-starved, impact-driven atmosphere? *I believe it must!* For example, we should make space for thoughtful, if risky, scientific proposals in grant applications, and in the evaluation process of research projects.

For a period of about a decade, the committees that allocated observing time on the Hubble Space Telescope were encouraged to give up to 10% of the time to proposals with a relatively lower probability of success but which promised potentially high return. A similar philosophy could, and I would argue should, be adopted more widely.

One problem with such a practice is that large committees are not inclined to approve risky programs. Efforts to reach consensus tend to converge to a mean. Such obstacles can be overcome if decisions are left to a small number of people or even to one person. In the case of Hubble, a pool of »director's discretionary time« on the telescope is available, for which anyone can directly apply. From this pool came such landmark observations as the »Hubble Deep Field«, one of the most detailed images of the Universe at large ever made.

Today, many telescopes on the ground and in space are devoted to addressing the profound outcome of another famous ›blunder‹. Einstein regretted his attempt to model a static cosmos using a repulsive-gravity force, which he introduced into his equations in the form of a ›cosmological constant‹. Since 1998, however, we have known that the cosmic expansion is accelerating, propelled by the repulsive force exerted on space-time by a smooth form of »dark energy.« So far most observations are consistent with this »dark energy« being precisely represented by Einstein's cosmological constant (which is a manifestation of the energy contained in the physical vacuum). What Einstein considered to be a »blunder« may thus have turned out not to be a blunder at all. In fact, it might have been an incredible insight on his part. Understanding the precise nature of this »dark energy« is one of the biggest challenges that physics is facing today.

To conclude, researchers must embrace blunders that come not from sloppiness, but from thinking outside the box. Those types of mistakes, which I dubbed »brilliant blunders« (Livio 2013), can be portals to discovery. Evaluation processes should allow for originality and creativity, even at the risk of committing the occasional blunders.

References

Bailes, M., Lyne, A.G., & Shemar, S.L. 1991. »A Planet Orbiting the Neutron Star PSR1829–10.« *Nature* 352: 311–313.

Livio, M. 2013. *Brilliant Blunders: From Darwin to Einstein*. New York: Simon & Schuster.

Lyne, A.G. & Bailes, M. 1992. »No Planet Orbiting PSR1829–10.« *Nature* 355, 213.

Redfield, R. 2012. »The CsCl/mass spectrometry data.« http://go.nature.com/bmb62d.

Tawfik, D.S. & Viola, R.E. 2011. »Arsenate Replacing Phosphate: Alternative Life Chemistries and Ion Promiscuity.« *Biochemistry* 50 (7): 1128–1134.

Thomson, W. 1867. »On Vortex Atoms.« *Proceedings of the Royal Society of Edinburgh* 6: 94–105.

Wolfe-Simon, F. et al. 2011. »A Bacterium That Can Grow by Using Arsenic Instead of Phosphorus.« *Science* 332: 1163–1166.

Discovering, Inventing, Contriving – What Are Scientists Actually Doing?

Gernot Münster

This contribution is based on a public talk, given at the workshop »Making Scientific Discoveries« in December 2019 in Münster, and addressing a general interested audience. How come that a physicist speaks about a topic, which touches various scientific fields and includes aspects of the philosophy of science? Originally, I intended to speak about discoveries in physics, but due to the unavailability of other speakers for the public talk, I was asked to fill in. I hope not to overdraw.

4.1 Introduction

In addition to the verb »discovering«, corresponding to the subject of this volume on making scientific discoveries, the title of this contribution contains the verbs »inventing« and »contriving«, and we shall discuss them later. To start with, I will just talk about »discovering« for simplicity.

Discoveries play a central role for the progress of science. To mention just a few in different fields:
- Exoplanets, planets orbiting other stars, discovered by M. Mayor and D. Queloz in 1995; awarded with the Nobel price in 2019
- The Higgs particle, discovered at CERN in 2012
- The Coelacanth fish (*Latimeria chalumnae*), discovered in 1938
- Penicillin, antibiotic effectiveness discovered by A. Fleming in 1928
- Entropy, discovered by R. Clausius in 1865
- Irrational numbers, discovered by Pythagoreens in the 5th century B.C.

In my contribution I will concentrate on discoveries in the natural sciences, and many examples will be from my own field, physics.

Now, »discovery« can in general either refer to the process of discovery or to the result of it. In the philosophy of science, the discussion of discovery has mostly been concerned with the process (see Schickore 2018). This includes topics like the »eureka moment«, creativity, the non-analysability of the process of discovery, abduction, phases of discovery, context of discovery versus context of justification, methods of discovery etc.

© BRILL MENTIS, 2022 | DOI:10.30965/9783957437044_005

This contribution will, however, deal with the results of discovery and their characteristics, although the title might suggest it differently. A central question to be discussed is this: which kinds of discoveries exist? We shall try to sketch a draft of a taxonomy of discoveries. This in itself might not be much more interesting than stamp collecting. But en route we will also have a look onto what actually makes up a scientific discovery, and we will discuss some related topics of the philosophy of science. And finally we shall make an attempt to specify what science is at all.

4.2   Discoveries

In his contribution to this volume, Kim Boström (2021) nicely analyses the discovery of the planet Neptune, predicted by U. Le Verrier on the basis of perturbations of Uranus' orbit, and observed by G. Galle 1846 in Berlin. The discovery of Neptune represents the first kind of discovery I would like to single out, namely individual *concrete material (real) objects/things*. Other examples of this kind are the metallic core of earth, the tomb of Tutankhamun, and the black hole in the Galactic Center.

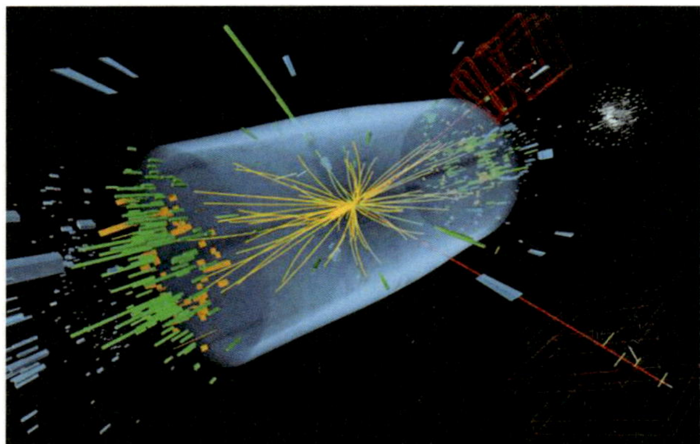

Figure 4.1   Higgs event recorded in 2012 by the CMS detector at the Large Hadron Collider in proton-proton collisions

Closely related to this kind of discoveries are *classes or types of such objects*. The Higgs particle, discovered experimentally at CERN in 2012, is an example. Each single Higgs particle is an individual concrete material object, but the discovery is about the Higgs particle as a type, characterised by its mass, spin, interactions and other properties. The same applies to exoplanets, and to Coelacanth fish (*Latimeria chalumnae*), which were believed to be extinct since the late Cretaceous period, and discovered 1938 by Marjorie Courtenay-Latimer.

Figure 4.2   Preserved specimen of *Latimeria chalumnae* in the Natural History Museum, Vienna, Austria

4.3   Theory-ladenness of Observation

The discoveries of the metallic core of earth or of the Higgs particle are good occasions for a digression about the so-called theory-ladenness of observation. Our knowledge about the structure of the inner of the earth and its chemical composition is to a large extent based on the study of the propagation of sound waves. To draw conclusions from the observational data about the material inside the earth presupposes a number of physical assumptions, which in turn have been tested in other circumstances. These assumptions form the so-called background theories. In the case of the Higgs particle the background theories include models from the theory of elementary particles and from the physics of detectors. Many other discoveries, e.g. of dark matter, gravitational lenses, etc., heavily rely on background theories. This kind of theory-ladenness is called »perceptual«.

Moreover, observations are not just made erratically, recording everything for possible later use. As Kant observed: »[...] it is undoubtedly certain that nothing of a purposive nature could ever be found through mere empirical groping without a guiding principle of what to search for.«[1] The choice of aspects and the focus of attention is taken according to theoretical conceptions. An instructive example is Brownian motion, discovered by R. Brown in

---

1   See Kant [1788] 1977, 141. The German original reads: »so ist wohl ungezweifelt gewiß, daß durch bloßes empirisches Herumtappen ohne ein leitendes Princip, wornach man zu suchen habe, nichts Zweckmäßiges jemals würde gefunden werden«.

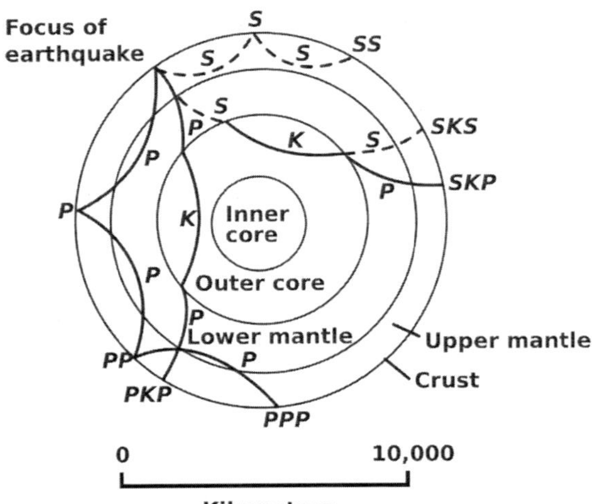

Figure 4.3
Paths of earthquake waves
inside the earth.

1827. Its explanation succeeded only after the theories of A. Einstein (1905) and
M. Smoluchowski (1906) motivated to measure the mean quadratic distance as
a function of time. The fact that observation is often guided by theoretical presuppositions is called the »searchlight theory of science« by K. Popper (1962).
Theoretical background plays a role in raising the awareness for the object
of discovery. It is often only against a theoretical background that the object
of discovery is regarded as new and scientifically interesting. As an example,
let me mention the physicist A. Goodspeed, who produced a photographic
picture with X-rays already in 1890, five years before W. Röntgen's great discovery by chance. Goodspeed, however, didn't pursue his observation, which
was inexplicable to him, and thus missed the opportunity for a fundamental
discovery.

The history of the discovery of the electron provides another case of theory-
ladenness. The discovery is attributed to J. J. Thomson, who in 1897 investigated
the deflection of cathode rays in electric and magnetic fields, and gave the
interpretation of the results in terms of the motion of charged massive particles. A comparable experiment was conducted by W. Kaufmann in Berlin, even
with higher precision. Kaufmann is, however, never referred to as discoverer of
the electron, because he didn't believe to have discovered a new particle. In his
opinion it wasn't the task of physicists to speculate about unobservable things.
Therefore he didn't state to have discovered a new type of particles, but just
reported that there was a certain ratio of electric charge to mass in what was
flowing in cathode rays.

Another kind of theory-ladenness of observation is called »semantic«. It
implies that the meaning of observational terms and statements is determined

through the network of associated theories. We shall not delve deeper into this important topic of the philosophy of science, but return to the discussion of objects of discovery.

4.4   Discoveries and Inventions

John Bardeen (1908–1991) is one of the few scientists who have been awarded the Nobel price twice, once 1956 for the discovery of the transistor (1947) together with W. H. Brattain and W. B. Shockley, and 1972 for the BCS theory of superconductivity (1957) together with L. N. Cooper and J. R. Shrieffer. Let us consider the transistor. The verb »to discover« is not appropriate here. As the word says, it figuratively indicates the removal of a cover from something that has already been in existence before.

In contrast, »to invent« means that something new is created or produced. Therefore, in the case of the transistor we rather have an example of an *invention*. Other examples are the laser (Th. Maiman, 1960) and the dynamo (W. von Siemens, 1866).

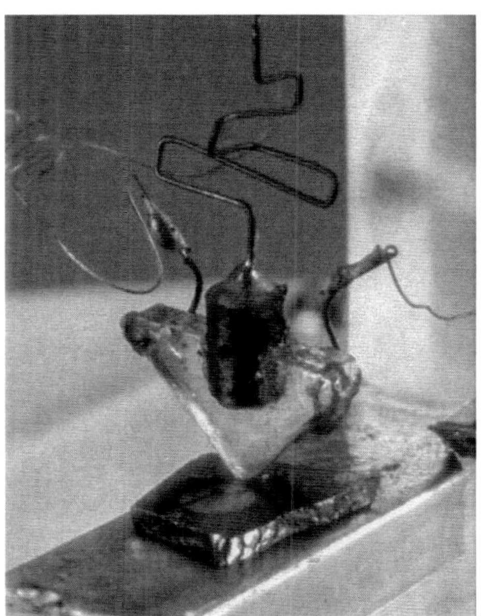

Figure 4.4
The first point-contact transistor, 1947

It might appear that relevance for applications is characteristic for inventions. This is, however, not necessarily so. There are counterexamples, one being the light mill (W. Crookes, 1873), which doesn't have any technical applications.

Figure 4.5
Light mill/Crookes Radiometer

Let us return to discoveries. Above we discussed classes of concrete material objects. »Class« and »type« are abstract concepts already. Can we make *discoveries of classes* per se, i.e., not only of the members of their respective classes? There are such cases where the objects have already been known before. An example is the discovery of the 32 crystal classes (M. Frankenheim 1826, J. Hessel 1830), which classify crystals according to the symmetry of their underlying crystal lattices. The different types of crystals were known, but not the crystal classes. Another example is the discovery of superconductors, which support electric currents without resistance. The substances (mercury, lead, etc.) were known before, but not the class of superconducting materials. This example shows a close relation to the corresponding properties: the classes are almost always defined by way of reference to certain properties of their members. Therefore, we should consider this kind of discovery rather as a *discovery of properties*.

This leads us to the next type: *discoveries of abstract non-material entities*. We will use the word »entity« in a very general way for objects, events, processes, properties etc.

Examples of *abstract non-material objects* that have been discovered in the natural sciences, are Planck's quantum of action (M. Planck, 1899), represented by Planck's constant h = $6.626... \times 10^{-34}$ Js, and the honey bee dance language (K. von Frisch, 1920, 1944/45; Nobel price 1973).

In the field of mathematics one would consider examples like the irrational numbers (Pythagoreens, 5th century B.C.), exemplified by the length of the diagonal in the unit square, exotic manifolds (spheres: J. Milnor, 1956; R^4: M. Freedman, 1982), or the Monster Group, the largest finite sporadic group with approximate 8×10^{53} elements (R. Griess, 1982). These examples, however, lead us to the next digression.

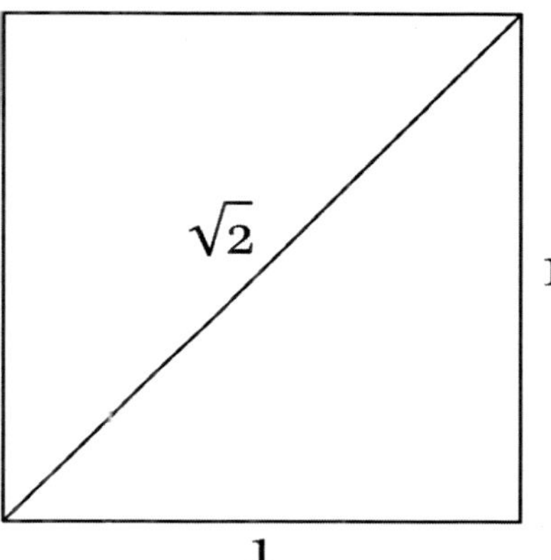

Figure 4.6
Irrational diagonal in the unit square

4.5   Realism

Are the irrational numbers or the Monster Group discoveries? Or are they rather inventions by mathematicians? This question is closely related to the question whether these objects are real. »Real« and »reality« are ambiguous concepts. Roughly speaking, they mean that something exists independently of human reasoning. In philosophy, discussions about the concept of reality are wide-ranging. Various types of realism are distinguished, two of them being ontological realism and epistemic realism. Ontological realism claims that there is a reality, which in its existence and constitution is independent of human experience, forms of thought and assumptions. Epistemic realism asserts that reality can be recognised partially. On the other hand, there are various versions of anti-realism like idealism, constructivism, etc. The controversy between realism and anti-realism fills libraries. In our context, scientific realism is of relevance (cf. Suhm 2005). Its ontological version asserts that the objects and properties which are postulated by an accepted scientific theory exist independently of human epistemic capability and linguistic descriptions.

Basically the quest of realism is about the question, what »real« and »existent« mean. In this context, I would also like to refer to the contribution of

Niko Strobach to this volume (Strobach 2021). The question whether universals (general concepts) are real or not is very old and comprises the »problem of universals« in the history of philosophy. Here the opponents are realism (Platonism) versus nominalism (anti-realism). The topic is very intricate, and I am reminded of Peter McLaughlin's aphorism »philosophy means, to make a problem out of a simple thing, and then to find no solution«.

Now let us consider the question of realism in the case of mathematics. Do numbers, functions, etc. exist independently from us humans? There are two main points of view. *Platonism* or mathematical realism, with G. Frege, K. Gödel, H. Putnam, S. Psillos being some of its proponents, claims the existence of these abstract objects, and considers mathematical objects as being discovered, as recent Fields Medalist P. Scholze emphasises. The opposite position is *constructivism or mathematical fictionalism* (other variants are logicism, formalism and intuitionism), which holds that mathematical objects are free creations of the human mind. Proponents of this view are e.g. L. Brouwer, P. Benacerraf and H. Field. The following quote by Leopold Kronecker (1823–1891) is well known: »God made the integers; all else is the work of man« (Weber 1893, 19). However, many mathematicians cannot uniquely be assigned to one of the two sides. P. J. Davis and R. Hersh write that »the typical working mathematician is a Platonist on weekdays and a formalist on Sundays« (Davis & Hersh 1981, 321).

To conclude, we have to leave the question open whether mathematical objects are discovered or invented.

4.6   Abstract Non-material Entities

We return to the consideration of abstract non-material entities outside mathematics. The next case is that of *events*. The observation of supernova SN 1987A and the detection of the gravitational waves produced by black hole mergers by LIGO in 2015 are examples of scientific discoveries of events.

In contrast to events, which are tied to specific short intervals in time, *processes* are not limited in this way. Examples of discovered processes are the blood circuit (W. Harvey, 1616), the oscillation of Cepheid stars (E. Pigott, J. Goodricke, 1748), the nuclear Bethe-Weizsäcker cycle (H. Bethe, 1939; C. F. v. Weizsäcker, 1937/8), which describes the transmutations of nuclei in the inner of the sun and of stars, and the urea cycle (H. Krebs, K. Henseleit, 1932).

Analogous to processes are *methods/procedures* like air liquefaction (Carl von Linde, 1895), scanning tunneling microscopy (G. Binnig, H. Rohrer, 1981, Nobel price 1986), radiocarbon dating (W. F. Libby, 1946, Nobel price 1960).

Examples from mathematics include the sieve of Erathostenes for finding prime numbers and Euclid's algorithm for computing the greatest common divisor of two integers. Methods and procedures, in natural sciences as well as in mathematics, should be counted as inventions and not as discoveries.

Abstract non-material entities that can be discovered also comprise *properties*, which we have mentioned above in connection with superconductivity. These include the structure of the benzene molecule (A. Kekulé, 1865), the constructability of the regular 17-gon (C. F. Gauß, at age of 18 in 1796) and the medical effectiveness of Penicillin (A. Fleming, 1928; Nobel price 1945).

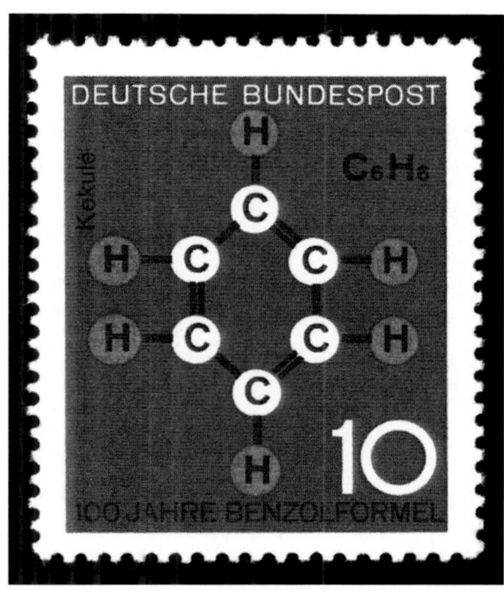

Figure 4.7
Structure of the benzene molecule,
German stamp 1964

Theoretical concepts also belong to the class of abstract non-material entities. In thermodynamics the central and important concept of entropy was introduced by R. Clausius in 1865. Entropy is an abstract variable that characterizes thermodynamic equilibrium states in addition to energy and temperature and is closely related to heat. Did he discover it or did he invent it? Is entropy something that exists in nature, or is it a pure human construction, that helps us understand phenomena involving heat? This question is analogous to the one about mathematical objects, and the answer depends on one's metaphysical point of view. One could also call entropy a concept that has been *contrived* (in German: »ausgedacht«, maybe »excogitated« fits better), meaning that something has been constructed mentally in the framework of a theoretical structure, and that doesn't exist as a real entity in nature. I would consider path integrals, introduced by R. Feynman in 1948, as contrived, too. They constitute an alternative formulation of quantum mechanics, established in a known theoretical framework, and mathematical equivalent to the standard formulations of wave mechanics or matrix mechanics.

Laws are more comprehensive than theoretical concepts. Examples are Kepler's laws (J. Kepler, *Astronomia Nova*, 1609, *Harmonices mundi*, 1619), Mendel's laws of inheritance (G. Mendel, 1865), the uncertainty relations (W. Heisenberg, 1927), and in mathematics the Four Squares Theorem about the representability of numbers as sums of four squares (J.-L. Lagrange, 1770), the law of quadratic reciprocity (conjectured by L. Euler, 1740; proven by C. F. Gauss, 1801). Also mathematical proofs, like Euclid's proof of the infinity of prime numbers can be counted as laws.

A level higher than laws are *theories* and their properties, and theoretical *models*. Here we have the General Theory of Relativity (A. Einstein, 1915), Quantum Mechanics (W. Heisenberg, 1925; E. Schrödinger, 1926), and string theory for example.

4.7   A Sketch of a Taxonomy of Discoveries

Let us summarise the previous paragraphs in the following table.

			Discovered	Invented	Contrived
Material	Concrete	Objects	x		
	Abstract	Classes of objects	x		
Abstract non-material entities	Spatially localized	Objects	x		
	Temporally localized	Events	x		
		Processes	x		
		Methods		x	
		Properties	x		
	Theory elements	Concepts			x
		Laws	x		x
		Theories/ models	x		x

This might be considered as a starting point for a taxonomy of scientific discoveries. As mentioned in the introduction, it is, however, not meant as the sole aim of my considerations. Looking at the different kinds of discovery, I will try to extract insights into what constitutes science at all. But for this purpose, I have to add another digression.

4.8   Sociological Aspects of Discoveries

What has so far not been taken into account in this contribution is the role of society for scientific discoveries. When I discover a mole in my garden, this is a discovery, but not a scientific one. Is it because it happens in a private territory and not in a laboratory? No: if I discovered a mini dinosaur, like the 80 kg *Moros intrepidus*, in my garden, this would certainly be considered to be a scientific discovery. Actually, *Moros intrepidus*, living in the Cretaceous period, has been found by paleontologists in 2013 and identified as a new species in 2019. What matters is the *relevance for the society or community*, in the case of scientific discoveries the scientific community. Consequently, before something can be considered to be a scientific discovery, it has to be *acknowledged* as such by the relevant community. Jan Michel has discussed this aspect in the context of declarative speech acts and institutional reality (Michel 2019, Michel 2021). In particular, a discovery which is not disclosed to anybody cannot be counted as a scientific discovery.

Figure 4.8   Mini dinosaur *Moros intrepidus*

Another aspect is the following. For something to be counted as a scientific discovery, it has to provide a *news value* for the corresponding community. And this will depend on which community is being considered. Think of the Qumran Caves Scrolls. The shepherd boy Mohammed Ad-Dib from the arabic tribe of Ta'amireh found these scrolls in 1947 by accident, when searching a lost goat in the region between Bethlehem and the Dead Sea. They are some of the oldest known biblical documents. They represent a discovery for us, but of course they were known to past communities. Similarly, the retrieval of the tomb of Tutankhamun was a discovery for archaeologists of the last century, but the ancient Egyptians knew it a long time before.

Conversely, if a ›discovery‹ is already known to the current scientific community, it will not be acknowledged as such. If some private scholar ›discovers‹ today, say, that the earth moves around the sun, it doesn't have a news value for us.

4.9   What is Science?

Above, I have collected different kinds and examples of scientific discoveries and discussed some related aspects. But what distinguishes them from discoveries in general? What is science?

We shall not be content with the simplest answer that is nonetheless heard sometimes, that »science is what scientists are doing«. A more sophisticated characterisation would be: »Science is the pursuit of objective knowledge«. There is even a definition given by the German Federal Constitutional Court, which is in the same spirit: »Scientific practice is everything, which according to its content and form is to be considered as serious, systematic attempt to seek the truth.«[2] But I don't find it sufficient enough, as it doesn't touch the role of background knowledge, community and relevance.

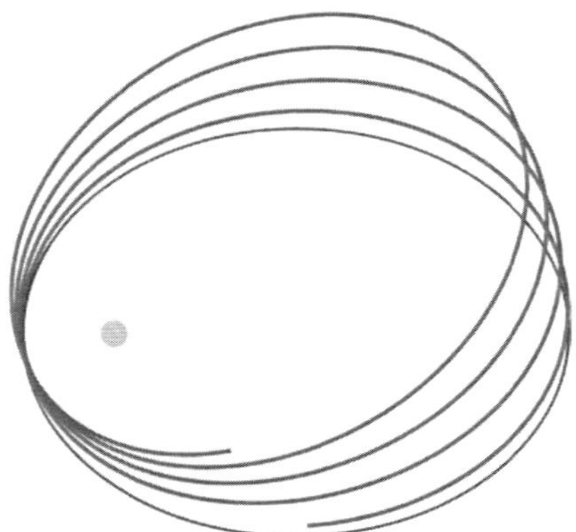

Figure 4.9
Perihelion precession of Mercury

---

2   »Wissenschaftliche Tätigkeit ist alles, was nach Inhalt und Form als ernsthafter planmäßiger Versuch zur Ermittlung der Wahrheit anzusehen ist.« Judgment of the Federal Constitutional Court of May 29, 1973 (BVerfGE 35, 79 CII) and January 11, 1994 (1 BvR 43c/87).

And what about »truth«? Are scientific theories or laws true? Consider Kepler's laws. Each planet is approximately moving on an elliptic orbit, but there are the disturbances from other planets, and there is the perihelion precession, predicted by Einstein's theory of gravity. Moreover, strictly speaking classical physics is superseded by quantum physics. Thus, although Kepler's laws provide a rather accurate and useful description of planetary motion, they are not exact, and they rest on an outdated ontology. Therefore, they cannot be called »true« in the proper sense of »truth«. In the same way, »truth« isn't an appropriate attribute of theories. Classical mechanics has been superseded by quantum mechanics, which in turn has been superseded by quantum electrodynamics, etc. These theories provide an approximate description of nature in a certain domain of applicability, but are not »true« in the proper sense. It appears more appropriate to speak of »rational acceptability« instead of »truth«.

In the philosophy of science, there is no general consensus about what science is. Different criteria have been proposed as characteristic features, e.g., falsifiability by K. Popper (1934), the potential for problem solving in the framework of a paradigm by T. S. Kuhn (1970), and systematicity by P. Hoyningen-Huene (2013).

Having the above examples and kinds of scientific discoveries in mind, and taking into account what has been said about news value, community, background knowledge, rational acceptability, and general relevance, and observing that news value, authenticity and significance must be testable empirically or theoretically, I propose to define:

Science is a collective enterprise with the aim to gain, in the light of recognised background knowledge, rationally acceptable, empirically or theoretically testable knowledge, whose relevance goes beyond the individual case, and to convey these insights to the community.

Acknowledgements

I would like to thank Jan Michel and Kim Boström for interesting and instructive discussions and for their extensive comments on a first draft of this contribution.

References

Boström, K.J. 2021. »Making Discoveries in Physics.« This volume.
Davis, P.J. & Hersh, R. 1981. *The Mathematical Experience*. Brighton: Harvester Press.
Hoyningen-Huene, P. 2013. *Systematicity: The Nature of Science*. Oxford: Oxford University Press.

Kant, I. [1788] 1977. »Über den Gebrauch teleologischer Prinzipien in der Philosophie.« In *Schriften zur Naturphilosophie*, edited by W. Weischedel, 137–170. Collected Works, vol. IX. 12th ed. Frankfurt: Suhrkamp.

Kuhn, T.S. 1970. *The Structure of Scientific Revolutions*. 2nd edition. Chicago: University of Chicago Press.

Michel, J.G. 2019. »How Are Species Discovered? Declarative Speech Acts in Biology.« *Grazer Philosophische Studien* 96 (3): 419–441.

Michel, J.G. 2021. »Toward a Philosophy of Scientific Discovery.« This volume.

Popper, K.R. 1934. *Logik der Forschung: Zur Erkenntnistheorie der modernen Naturwissenschaften*. Wien: Springer. Online source: https://monoskop.org/images/8/83/Popper_Karl_Logik_der_Forschung_1935.pdf. Translated into English as *The Logic of Scientific Discovery*. London, 1959.

Popper, K. 1962. *Conjectures and Refutations: The Growth of Scientific Knowledge*. New York: Basic Books.

Schickore, J. 2018. »Scientific Discovery.« In *The Stanford Encyclopedia of Philosophy* (Summer 2018 Edition), edited by E.N. Zalta. https://plato.stanford.edu/archives/sum2018/entries/scientific-discovery/.

Strobach, N. 2021. »How Much Realism Does the Word ›Discovery‹ Presuppose?« This volume.

Suhm, C. 2005. *Wissenschaftlicher Realismus*. Heusenstamm: Ontos/De Gruyter.

Weber, H. 1893. »Leopold Kronecker.« In *Jahresbericht der Deutschen Mathematiker-Vereinigung*, vol. 2, 5–31. Berlin.

Figure Sources

Fig. 4.1: © CERN, 2012

Fig. 4.2: https://commons.wikimedia.org/wiki/File: Latimeria_Chalumnae_-_Coelacanth_-_NHMW.jpg

Fig. 4.3: https://commons.wikimedia.org/wiki/File:Earthquake_wave_paths.svg

Fig. 4.4: https://www.computerhistory.org/revolution/digital-logic/12/273/1355, © 2006-2007 Alcatel-Lucent

Fig.4.5: https://commons.wikimedia.org/wiki/File:Radiometer_Crookes_-_Oberweißbach-Thüringen.JPG

Fig. 4.6: own drawing

Fig. 4.7: https://commons.wikimedia.org/wiki/File: Stamps_of_Germany_(BRD)_1964,_MiNr_440.jpg

Fig. 4.8: https://dinosaurpictures.org/Moros-pictures, © Jorge Gonzalez

Fig. 4.9: https://commons.wikimedia.org/wiki/File:Relativistic_precession.svg

How Much Realism Does the Word ›Discovery‹ Presuppose?

Niko Strobach

5.1   Introduction[1]

It is best to start with a nostalgic methodological remark. Once upon a time, there was a linguistic turn in philosophy. Philosophers saw a point in looking at words. They followed the advice of Willard Van Orman Quine to reformulate questions in terms of »semantic ascent«.[2] They knew: Just as boots are made for walking, quotation marks are made for mentioning. They knew that scare quotes are a bad thing, because they create the illusion that one can escape responsibility for one's own use of words. They remembered Rudolf Carnap's idea of the logical syntax of a word: Find out the shortest sentence in which the word may occur and make sense of it.[3] The useful philosophical technique of looking at words should not be forgot. So, in this paper, the focus shall be on the word »discovery«, and on similar words, like its basic component, the verb »to discover«.

If one pays attention to the formation of the two words, they turn out to contain an image (or even a metaphor): lifting or removing some opaque cover, such that what used to be invisible underneath becomes visible. This observation provokes a question:

> »Is there a link between the word ›discovery‹ and the opposition between realism and anti- realism in the philosophy of science?«

---

1   I am grateful to Jan Michel for inviting a contribution on this topic. As far as I am aware, it does not connect to an ongoing debate. Thus, the use of literature is entirely restricted to the purpose of reference. The paper originated as a talk, and I saw no point in artificially avoiding the first person singular. This might even suit the text's being both opinionated and tentative. Thanks go to all participants of the discussion in Münster on 3rd December 2019, in particular to Mitch Green, Christian Tapp, and to Jörg Friedrich (also for valuable information on Heidegger; he may disagree with my interpretation). Further thanks for additional information and for challenging remarks (largely unmet in what follows) go to Jan Michel and Peter Rohs.

2   Quine 1960, 270–276.

3   Carnap 1931, 221f.

Presupposing that the image exists, one may imagine different answers:

> (Answer 1) »No, there isn't. We should forget about the cloudy terms ›realism‹ and ›anti-realism‹.«

> (Answer 2) »No, there isn't. We all know well enough what ›realism‹ and ›anti-realism‹ means. But the image, which is contained in the word ›discovery‹, is irrelevant to the use of the word ›discovery‹ in the philosophy of science. Poetry is not an issue here.«

> (Answer 3) »No, there isn't. The image is relevant. But the realist and the anti-realist (whom we know well enough) can both include it in their use of the word ›discovery‹, each in his own way.«

> (Answer 4) »Yes, there is. It is even a strong link: Because of the image contained in it, ›discovery‹ is a word that the realist can seriously use, while the serious anti-realist should refrain from using it (he might use words like ›invention‹ instead).«

The answers are pairwise incompatible. I do not claim that the list is exhaustive. I would like to persuade the reader to take answer 4 seriously. In the end, however, a more differentiated position will emerge as the most plausible one: There are occasions, on which an anti-realist may seriously use the word »discovery«. But, apart from observable entities hitherto unnoticed, they concern intra-theoretical connections. And they will turn out to be occasions for using the word not in any special sense, but in its basic sense in which the *realist* uses it anyway. This shows how much realism the word »discovery« presupposes: quite a bit.

I shall start by roughly characterizing realism and anti-realism (section 2). This will, I hope, make answer 1 implausible enough. I'm not neutral, I am a realist. I shall then turn to metaphors (section 3) and provide some information on words in different languages (section 4). This will be a part of my efforts to make answer 4 look plausible. In the course of that, I will try to make answer 2 look implausible – no K.O. is to be expected, though. I will then try to clarify the logical syntax of »discovery« and »to discover« (section 5). I shall distinguish between an objectual and a propositional use. Occurrences of »to discover« followed by a direct object (»America«, »the atom«) should be distinguished from occurrences followed by a that-clause. This will touch upon answer 3. It will turn out that realists might have some affinity to the direct object construction, and anti-realists to that-clauses, and this is telling. But things will not quite be so simple: Language is flexible, and only some *onto-logical* considerations will clarify the matter (section 6). Finally, a challenge to the realist must be addressed: Can *he* ever use the verb »to discover« in a true sentence if it is an achievement verb (and even veridical)? The answer is: It is, and he can (section 7).

I do not hope to persuade anyone who is not a realist already. I am content if I can make the following question seem relevant to realists and anti-realists alike:

> »What *is it* that, in a given instance of use, the word ›discovery‹ is applied to?«

To a reader whom I cannot even convince of this, the data on words in section 4 might still be of some use.

5.2   Realism and Anti-realism

What might the word »realism« mean? And how about its opposite, »anti-realism«? A special problem adds to the well-known general difficulties of definition[4] here: In *my* opinion, the verb »to discover« would come in handy for the task. But using it would be question-begging.

A rough outline: Realists believe in the theory-independent existence of the objects a true scientific theory is about. They believe in the theory-independent existence of facts.[5] They believe in the theory-independence of truth: Truth values of many sentences depend on how reality itself is; even though, trivially, they are formulated in the vocabulary of some theory, they are not made true *by* that theory. Realists believe that the pursuit of truth is valuable. They are optimistic that we have already found out many truths about how reality itself is. For instance, we are in a position to give a true answer to what was a mystery question to David Hume:

> »Who will assert that he can give the ultimate reason, why milk or bread is proper nourishment for a man [...]?«[6]

A unifying characterisation of *anti*-realists is hard to find. I shall not try to formulate one. This is not an anti-anti-realist paper. Otherwise I would have to be clearer whom I oppose.

---

4   There are many attempts at defining the terms. A good collection of texts on the topic is Papineau 1996. The editor's introduction (ibid., 1–20) is particularly useful.

5   A little more will be said about facts in section 6. But it is best to take the word »fact« at face value and not think about it too much while reading through this paper. Even in a philosophy text, it is impossible to analyse every word at the same time at which one is using it.

6   Hume [1772] 1975, 28.

Many anti-realists, not all of them, say that the objects a scientific theory is about (for instance, atoms) are constructions; that, therefore, they exist relative to the theory; and that scientists create theory-relative facts. They believe that sentences are true relative to the theory whose vocabulary is used to formulate them and that their truth value does not depend on reality itself, because there is no such thing. They think that the pursuit of truth, *as the realist conceives of it*, is neither worthwhile nor successful, because no such truth exists. They may affirm that there are theory-relative truths, which are even objects of (theory-relative) knowledge.[7] In my view, Ludwik Fleck was a very intelligent advocate of this position.[8]

There are also more moderate anti-realists – followers of Bas van Fraassen –, who do not mind reality itself, but maintain that scientists should not care about which objects or facts it contains; rather, they should strive for good models which are adequate to the empirical data that the observation of reality itself provides them with.[9] In my view, the progress that van Fraassen made by adding his position as an option to other anti-realisms is precisely that he showed how one can be a traditionalist about truth (»atoms exist« is true if and only if atoms exist) and accept reality itself without ceasing to be an anti-realist. As van Fraassen himself suggests,[10] this is progress compared to authors of the French Instrumentalist tradition who may be interpreted as holding that an emphasis on the instrumental character of theories and theoretical entities entails non-traditionalism about truth.

I have decided to divide the camps by making use of the phrase »reality itself«. I am aware that constructivists may claim to be realists, namely realists of self-made reality.[11] But »self-made reality« means about the opposite of »reality itself«.

When I use the phrase »*my* anti-realist« in what follows, it may happen that every self-confessed anti-realist, who happens to read this paper, will lean back and say that he is not identical with the stooge designated by that phrase. It is not my business to make sure that he would be right to do so.

---

7   Fleck [1935] 1980, 70: »Certainly, there is a lot we are able to know.« (»Wir können sicher vieles wissen.«)

8   There are less intelligent advocates whom I need not mention. A criterion for finding out whether someone adheres to this position might be his reaction to the question »Did Ramses II die of tuberculosis?« Cf. Strobach 2011 on the question how far removed from realism Fleck's position can be intelligibly reconstructed.

9   Van Fraassen 1980.

10   Ibid., 2.

11   Once more, Fleck [1935] 1980 is the *locus classicus* for this claim.

5.3　Words and Metaphors

As I said, I think that looking at words is worthwhile. It may happen that the intelligible use of a single word presupposes a certain background, which is invoked when – as Ludwig Wittgenstein put it – language games are played with it.[12]

How does a word in use invoke a background? There are many ways in which it can achieve this. One of them is by activating an image, which is somehow inherent in the word. Often, this comes together with *transferring* the word (so to say): extending its application to some realm, which is different from the realm of application it originally had – maybe more abstract, so it was difficult to talk about. Should one say, then, that the use of the word in the new area of application is metaphorical or that it is a metaphor in that context? That seems plausible. A problem is that there are many theories of metaphor by philosophers of language,[13] which agree on only one point: that this is *not* how metaphors work. My solution for the present paper is: don't care. I will use the word »metaphor« in an unphilosophical sense, as, for instance Guy Deutscher does, in the great chapter on metaphors in his great book *The Unfolding of Language*.[14] However, the *transfer* dimension, which gives metaphor its name, is not too relevant right now. The dimension of figurative speech or imagery is.

Nevertheless, there is something useful about using the word »metaphor«: Doing so enables me to use the well-established technical term »dead metaphor«. The chapter in Deutscher's book has the – nicely metaphorical – title »A reef of dead metaphors«, since this is – metaphorically put, of course – the ground on which speakers of today's languages stand. A dead metaphor is an expression, which, in its current main area of use, used to be a metaphor, but no one is aware of this when using the word today. So, strictly speaking, a dead metaphor is not a metaphor, just as a dead person is not a person or an ex-friend is not a friend.

This suggests reformulating answer (2) as

> (2*) We can (so to say) excavate the image, which is contained in the word »discovery«. But while we talk about science, the word »discovery« evokes no image at all. It is a wholly abstract word. The metaphor it used to be is stone dead.

12　Wittgenstein [1953] 1988, for instance §7, §23.

13　For an overview cf. Ortony 1993. An impression of the range of philosophical theories of metaphor may be gained from Black 1954, Searle 1977, Davidson 1978 and Glucksberg & Keyzar 1993 – to mention only a few.

14　Deutscher 2005, 115–143.

But *is* »discovery« a dead metaphor in this context? A disadvantage of the term »dead metaphor« is that it sounds pretty digital: either dead or alive, nothing in between. As against this, it seems to me that, when it comes to metaphors, there is a great range between being stone dead and being a very lively surprising poetic innovation. Thus, a lot of words seem to be *sleeping* metaphors. You do not pay any attention to it, until the word clashes with another sleeping metaphor or some other odd context and produces an embarrassing *catachresis* or image break.[15] There cannot be an image break without an image. So the metaphor had not been dead after all, the metaphor had just been asleep.

One point I would like to make in favor of answer (4) – »discovery« is a word exclusively for realists – is that, even in talk about science, the word »discovery« contains at least a metaphor that is asleep and, thus, not dead. How does one find out? Not by argument, but rather by exposure to data.

5.4   Some Data About Words

1) Deutscher writes:[16] »Dis-cover initially meant ›remove the cover from‹. In the seventeenth century, it could still be used in this physical sense: ›if the house be discovered by tempest, the tenant must in convenient time repair it.‹«

2) There is a German verb, which precisely translates »to discover«: »entdecken«. Also the metaphor is perfectly parallel. So is the word formation, but, of course, from different components (»ent« and »decken«).

3) The contrast between »to discover« and »to invent« is mirrored one-by-one by the contrast between »entdecken« and »erfinden«, where, what you invent, has not been there before. The metaphor »finden« (»find«) in »erfinden« is modified by the prefix »er«:[17] The object of »erfinden« has not been there before; it is created by the activity. This is markedly different from »finden«.

4) A book of Chinese stories for beginners contains the sentence: »Shén Nóng fā míng le chá«[18] – »Shen Nong discovered the tea.« The bi-syllabic Verb »fā míng« (发明) means »to discover« (or: »to invent«). But literally it means

---

15   Black (1954) uses the word in a different sense.
16   Deutscher 2005, 125.
17   Pfeifer 1995, 292.
18   Hornfeck & Ma 2019, 146 (back cover).

»to send light« or »to shed light on«. A different metaphor. Translation is compromise.

5) Biblical Hebrew has a verbal root גלה (»glh«), which, in its literal use, corresponds to the old English meaning of »to discover«. Metaphorically, it works just like »to discover«, also in constructions to be rendered by »discovering something to someone«.[19]

6) Among the ancient Greek words, which may, sometimes, be translated by »to discover«, is the verb »εὑρίσκειν« (»heuriskein«), »to find«.[20] Indeed, one iconic scene of discovery is Archimedes shouting »ηὕρηκα« (»Heureka!« – »I've found it«), when he discovered *that* any floating object displaces its own weight of fluid (we shall come back to »that«-clauses).

The metaphor of uncovering appears in several compound verbs with the verb »καλύπτειν« (»kalyptein«) – »to conceal«[21] – as its main component, among them »ἀποκαλύπτειν« (»apokalyptein«), from which »apocalypse« in the sense of »revelation«.[22]

A complicated case is the verb »ἀληθεύειν« (»alêtheuein«), together with the corresponding noun »ἀλήθεια« (»alêtheia«) and adjective »ἀληθής« (»alêthês«). The standard translations are »speak truth«, »truth« and »true« respectively. In most cases, the translation is straightforward and no idea of discovery intrudes. It is, however, true that »ἀ-ληθής« is composed like »unconcealed«.[23] So does »ἀ-ληθεύειν« have »to dis-cover« as a deeper meaning, and is »dis-covery« a deeper meaning of »ἀ-λήθε.α«? Hard to tell (but cf. 9).

7) Scholars think that a verbal root *wer must have existed in Proto-Indo-European (spoken somewhere in the Eurasian Steppe some 5500 years ago). Its meaning: to cover. Intensifying it by the prefix *op (roughly »onto«, from which »auf«, »up«, Greek »ἐπί«/»epi«), yielded a verb *op-wer-yo, which, over a long time, became Latin »operire«; which, when worn out a little, was once more intensified to »comoperire«, which, the »m« being rubbed off, became »co(o)perire« – »to cover up«. No antonym was derived from it in *classical* Latin. Rather, verbs like »denudare« (»denude«) and »detegere« (from which »detective«) and »invenire« (from which the Latin word

---

19   Gesenius 1905, 125.
20   Liddell & Scott 1940, 729, lists »find out, discover« as meaning II.
21   Ibid., 871.
22   Ibid., 201.
23   Ibid., 64.

»inventio«)[24] were used in the sense of »to discover«. Only *medieval* Latin brought about the antonym of »co(o)perire«: »disco(o)perire«, which led to Old French »descouvrir« (modern French: »découvrir«) and, around 1300, to Old English »discoveren«.[25]

8) How about »entdecken«? The Indo-European root *teg must have meant »to cover«, too. Latin »tegere« (remember »detective«), German »Dach« and English »deck«, as well as »to thatch« go back to it. In the course of a 3500 years (or so), *teg led to Old High German »thecken«, which became »decken«. A descendant of the Indo-European root *ant (from which, *via* ancient Greek, all the words beginning with the prefix »anti-«), allows for the formation of the antonym »intthecken« (Old High German), which becomes »entdecken« in modern German.[26]

9) Digression: »Entdecken« in German has a homophonic counterpart in Heideggerian,[27] conveniently notated as »entdecken[H]« (Heideggerian is a language which is parasitic on German). Martin Heidegger's treatment of »entdecken[H]« goes along with a deeper meaning thesis he holds concerning »ἀλήθεια«.[28] Mimicking Heideggerian by Henglish, which relates to English as Heideggerian does to German, one can say that Heidegger was up to equating »to dis-cover[H]« with »to be true[H]«: X is true[H] if and only if it does a job of dis-covering[H]. There are two very different subcases depending on how »X« is specified: (1) *Existence*[H] (»Dasein[H]«) is true[H] if and only if it does a job of dis-covering[H], such that dis-covering[H] is a mode-of-being[H] of existence[H].[29] (2) A *statement* is true[H] if and only if it does a job of dis-covering[H], i.e. by

---

24   The relation of »invenire« and »inventio« is a bit complicated: The verb may be used for any kind of finding out and is therefore not synonymous with contemporary English »to invent« (a new kind of machine, a story); in the time of classical Latin, Cicero, in his *De inventione.* uses the word »inventio« for finding or imagining a suitable theme for a speech or for the ability to do so (as late as 1720, Johann Sebastian Bach transfers this use of »inventio« to suitable themes in pieces of music). The *inventio medii* in medieval logic is about finding a middle term, not about inventing one.

25   Cf. the entry »discover« in Douglas Harper's *Online Etymology Dictionary* (www.etymon-line.com). For the points on Latin cf. also Walde 1906, 36f., 431. The OED entry »discover« provides some very old instances among which the case of revealing someone's identity, which was to be kept secret, is relatively prominent.

26   Pfeifer 1995, 207 & 287.

27   Heidegger [1927] 1986, 212–230 (§ 44).

28   Ibid., 229.

29   Ibid., 220: »Wahrsein als entdeckend-sein ist eine Seinsweise des Daseins [...] Primär ›wahr‹, das heißt entdeckend ist das Dasein.«

dis-covering[H] being[H] to existence[H] at a certain time.[30] To sum up: Heidegger was sensitive to the metaphor in »entdecken« (»to discover«), which is in focus in the present paper. But »to dis-cover[H]« is so different from »to discover« that it is of no use to the aims of the present paper.

10) Spanish seems to contain surprisingly many words, which may sometimes be translated by »to discover«: »detectar«, »revelar«, »encontrar«, and, finally, »descubrir«, from which a noun is derived, just as »découverte« is derived from »découvrir« and »discovery« from »to discover«: »descubrimiento«. I'm mentioning it because of the second iconic scene of discovery, which is referred to by the words »el descubrimiento de America« – »the discovery of America«.[31]

11) It seems that the usage of »discovery« which has nothing to do with roofs, tempests or knights lodging incognito, was at least promoted by, precisely, the discovery of America, as references in the *Oxford English Dictionary* (OED) from the 16th century show: »the discouerie of Cathay« (1554), »discouerie of the West Indies« (1603) (but also »discouery of [...] medicines« (1590)). In general, a maritime context with its ambivalent and gradually changing visual clues of what lies ahead is particularly prone to this usage. It partly overlaps with a military use of »discovery« in the sense of »reconnaissance« (Shakespeare: »diligent discouery«).

12) The sentence »Columbus discovered America« is widely considered to be a true English sentence – Vikings aside, and (more importantly) in spite of Amerigo Vespucci. This yields two interesting observations:

> (a) The verb »to discover« may be used for describing a situation *without any surprise.*

30　Ibid., 218: »Die Aussage *ist wahr*, bedeutet: sie entdeckt das Seiende an ihm selbst.« Ibid., 227: »Die Gesetze Newtons [zeigen] das Seiende [...] entdeckend auf [...]«.

31　These data on Spanish were gathered from Glosbe (glosbe.com). I do not claim that Columbus himself used any of these words for anything he did – probably he did not. As Jan Michel informed me, it seems that some authors classify »descobrir« and »descobrimento« as 15th century neologisms in Portuguese which emerged because of Portuguese maritime activities (Gibson 2014, 7). Given the development sketched in 7) above, I am sceptical about this, but, of course, a word may have been in the air in several Romance languages and have acquired particular importance and new paradigmatic instances of use when long distance sailing entered a new phase.

Columbus was not surprised, because he had set off to find something; and: because it never occurred to him that he had found a new continent. Which adds observation ...

> (b) Sometimes the content of the verb »to discover« is ascribed externally and in retrospect.

Furthermore, the verb »to discover« need not refer to a sudden or short event. The discovery of the atom took up the whole of the 19th century. Note that the discovery of the atom must not be confused with the invention of the *concept* ATOM, which was achieved, presumably, by Democritus and maybe on one particular lucky day for science, after which the concept underwent some transformations during the 20th century. As far as this case is concerned, I think it is apt to speak of the *invention* of the concept ATOM (though »introduction« or »formation« would be more usual), as the concept did not exist before Democritus' lucky day. The idea of using the word »discovery« at all in such cases is, I think, psychological: It felt as if the concept had been lying there, was already on the theorist's mind, when he became aware of it, as if some god had placed it there shortly before for a kind of intra-mental Easter egg hunt[32] – instead of being mind-crafted by him. Shocking as it may sound, I do not count this as a *serious* use of the word »to discover«. To sum up: Democritus did not discover any atoms. Neither did he discover the concept ATOM or atom theory, because concepts or theories are *invented* (however subconsciously), not discovered. By the way: If John Dalton and other 19th century chemists discovered the atom (i.e. discovered that atoms exist), then entities, which are too small to be directly observable, may yet be discovered. Seriously. This does not overstretch the core image in the word »to discover«. In order to be discovered (strictly and non-metaphorically), atoms do not have to be large enough to be seen crawling around once a blanket is pulled away.

5.5 Some Logical Syntax

The linguistic data may make the following questions appear at least motivated:
(a) How much realism in the philosophy of science must someone presuppose if he uses words like »discovery« or »to discover«?

---

32 Whatever Democritus may have thought about this, the idea of a God placing a thought into someone's mind or heart – φρήν is difficult to translate – was not alien to the ancient Greeks, cf. Homer's *Odyssey* 19,10; 21,3.

(b)   Can an antirealist use them at all without sounding weird?
(c)   Can, whatever is discovered, be called a construction?
Moreover, the answer »quite a bit« to (a) and the answer »no« to both (b) and
(c) might at least not look absurd by now (this supports at least considering
answer (4) from the introduction seriously).

Two details of the linguistic observations can be associated with the two
iconic scenes of discovery. The verb »to discover« may take a noun as its direct
object:

> (C1) »Columbus discovered America«.

Or it may be followed by a »that«-clause:

> (A) »Archimedes discovered that any floating object displaces its own weight of
> fluid«.

Sentence (C1) is paradigmatic. It refers to the second paradigmatic scene of discovery. It is close to the original literal use of »to discover«. It is not an instance
of it, though, because America was not under a blanket. The metaphor is in full
force. It is simple to grasp, for America is an (albeit large) concrete individual
material (mainly) unscattered object.[33] This nicely matches the realist's ideas:
What is discovered by science has been lying there, waiting for discovery, is
unchanged by discovery, but seen only after it.

In contrast, it may seem that the metaphor is somehow muted or downplayed by the »that«-clause construction in sentence (A). One may ask: What
is the object of discovery here? Is there any? Is not, what is discovered, just, *that*
a certain sentence, formulated in the language of a theory, is true? (»Weight«
is a theoretical term.) There was no *thing* there waiting to be discovered. This
matches my anti-realist's ideas.

This might be seen as support for answer 3 (»The image is relevant. But the
realist and the anti-realist can both include it in their use of the word ›discovery‹, each in his own way«): Let the realist have the direct object construction,
while the anti-realist construes »to discover« with »that«-clauses. But that is
not, in fact, answer 3. Answer 3 implies that both sides take the image seriously. That, however, is precisely not the case if the »that«-clause-construction
mutes the image, but in that case the image is downplayed. One would rather
be back at answer 2, but restricted to »that«-clauses: syntax kills the metaphor.

---

33   Whatever that means for the date of the discovery of America: In 1492, Columbus landed
      on some Caribbean island. Only in 1504 did he reach Honduras.

However, it is not even so clear that the »that«-clause-construction in fact mutes the image. While logico-grammatical considerations help us see what a controversy over the legitimate use of »to discover« would be about, solely relying on them would be making too much of them. Ontological considerations cannot be avoided.

5.6   Some Ontology

Here is an ontological question:

>»What is it that, in a given instance of use, the word ›discovery‹ is applied to?«

(1) Consider the following sentence:

(P1) »Priestley discovered a method for obtaining Oxygene.«

This is a direct object construction, but what kind of entity is a method?[34] Was it lying there, waiting to be discovered? If methods or techniques exist, a realist might react by declaring (P1) as a misuse of »to discover« and recommend replacing it by »to invent«.

(2) Certainly, a realist would like to say things like

(P2) »Priestley discovered Oxygene.«

(In line with Columbus, it must be Priestley – Lavoisier corresponds to Amerigo; one might say that Lavoisier discovered *that* Priestley had discovered Oxygene). However, apart from surface grammar, the word »Oxygene« does not quite work like the word »America«. The word »Oxygene« does not refer to one individual unscattered material object. Can the realist cope with this? He might say that »Oxygene« is a name of something, maybe a natural kind, which is not a material object; maybe a *scattered* material object.

---

34   During the discussion, Mitch Green suggested that this kind of talk about methods could be paraphrased away by »Priestley discovered how to make oxygene« etc. This is a third kind of construction, in which »to discover« occurs. I regard it as one, in which the metaphor is lost for lack of an object of discovery (the same does not apply to »that«-clauses, if they are names of facts). So the suggestion seems to me to amount to a restricted version of answer 2 – restricted to »how to«-cases.

(3) No less a logician than Frege maintained that there is no *logical* difference between the two grammatical constructions: Never mind *grammatical* objects, which are *words*. What matters is that, (onto-)logically speaking, »that«-clauses are names of objects (Gegenstände), too, namely names of *facts*.[35] (Onto-)logically speaking, both constructions are object (Gegenstand) talk.

Of course, facts are highly contentious among ontologists. One may refuse to admit facts to one's ontology and deny that a »that«-clause refers to any entity; and if there are no facts in the first place, they can, of course, not be discovered (entities of other kinds can). One may accept facts, but stress that, being immaterial, they must inhabit some realm of their own (Frege's own view). One may object to Frege that he exaggerates their subject-independence, while they are really all produced by minds at certain times (not *my* view). One may find *this* implausible and admit them as immaterial, neither necessarily mind-crafted nor necessarily timeless objects to one's ontology without reserving a special realm to them (why not?). Etc. etc. It seems to me that philosophers of science are relatively relaxed about admitting facts to their ontology and of considering them as candidates for objects of discovery as well as of knowledge.

(4) At first sight, it seems that every direct-object-construction can easily be turned into a »that«-clause, for instance:

(C2) »Columbus discovered that America exists«.

However, this sounds a bit odd. Maybe this is good, because that provokes a distinction: Yes, he did, that's what he discovered – and not by any theoretical activity but by actually sailing there. But he never got it that he had discovered a new continent (we shall come back to the point). Yet, the move from the direct object construction to the »that«-clause construction can be useful. The discovery of *which* object is referred to by the phrase »the discovery of the atom«? Hard to say. Better take this as an abbreviation for »the discovery that atoms exist«.

All this is no refutation of answer 3 (»The image is relevant. But the realist and the anti-realist can both include it in their use of the word ›discovery‹, each in his own way«). An anti-realist may welcome talk about facts, as long as they are theory relative, for instance. The real value of distinguishing

---

35　　»That«-clauses are names of thoughts (Gedanken), which may be true or false (Frege 1892, 37; Frege 1918, 60). »To discover« is veridical (cf. section 7). So in its context, »that«-clauses will always be names of *true* thoughts, and Frege defines »fact« as »true thought« (Frege 1918, 74).

between the two constructions lies somewhere else. The deeper reason, why there is some affinity of the »that«-clause construction and anti-realism, is this: »that«-clauses are well-suited for describing how intra-theoretical connections are discovered. It would be odd to say that these are invented. They are indeed there, waiting to be grasped. Let us focus on this case.

Given a certain theory, I notice that it commits me to affirming a certain sentence.[36] I had not been aware of that before. Maybe the sentence had occurred to me, but not the commitment to it.

Here is a third iconic scene of discovery. A few months *after* publishing his foundational article on Special Relativity »Zur Elektrodynamik bewegter Körper«, in 1905, Albert Einstein wrote in a letter to Conrad Habicht:[37]

> »A consequence of the study on electrodynamics did cross my mind. Namely, the relativity principle, in association with Maxwell's fundamental equations, requires that the mass be a direct measure of the energy contained in a body [...] The consideration is amusing and seductive; but for all I know, God Almighty might be laughing at the whole matter and might have been leading me around by the nose.«

The anti-realist might get a point in favor of answer (3) out of this. Remember that answer (3) was that both the realist and the anti-realist can exploit the image in the word »discovery«, *but each in his own way*. Accordingly, my anti-realist might say:

> »Indeed, there is something to the image of uncovering. *You* like it, I know the way *you* talk. But we must be clear about the *object of discovery*: What is discovered is no *theory independent* fact of the equivalence of mass and energy that was waiting for dis-covery – there is no such fact. What was waiting for discovery (*after* Einstein had invented Special Relativity) was the theory relative fact that $E = mc^2$.«

I think this is about the best point that can be made for answer 3. But is it convincing? I do not think so. For my anti-realist is misdescribing his object of discovery. What Einstein discovered was that »$E = mc^2$« follows from Special

---

36   In the terminology of Fleck (1935, 34): My thought-style forces me to realize a certain necessary association (»zwangsmäßige Kopplung«).

37   Einstein 1994, document no. 28, translation in the English translation supplement. The original German text is: »Eine Konsequenz der elektrodynamischen Arbeit ist mir noch in den Sinn gekommen. Das Relativitätsprinzip im Zusammenhang mit den Maxwell'schen Grundgleichungen verlangt nämlich, dass die Masse direkt ein Maß für die im Körper enthaltene Energie ist. Die Überlegung ist lustig und bestechend; aber ob der Herrgott nicht darüber lacht und mich an der Nase herumgeführt hat, das kann ich nicht wissen.«

Relativity. But it is not a theory relative fact that »E = mc²« *follows* from Special Relativity. It is a *plain* fact, a realist's fact. After checking his calculation, Einstein could be sure that God Almighty did not fool him about *that*. Here is another plain fact: the fact that it *follows from phlogiston theory* that the gas, which Priestley isolated, was phlogiston-reduced elementary air. Let us make the assumption that Priestley discovered that fact. Is this assumption synonymous with

> (P3) »Priestley discovered the fact that the gas he isolated was phlogiston-reduced elementary air«?

No. I think that this consideration supports answer 4: The typical cases for which the anti-realist might use the word »discovery«, if he respects the image contained in it, turn out to be cases of discovery of *real* intra-theoretical connections, not cases of discovery of theory relative facts. But then, he would be using the word in the *realist's* way, not in any special way of his own. He would respect the image in it as a *realist* image. He would admit that the word presupposes some realism – and accept this for intra-theoretical connections. If, on the other hand, he applies the word »discovery« to genuine *theory dependent* facts (in contrast to theory independent facts *about* some theory), he cannot respect the image contained in the word.

Maybe my anti-realist is upset and says *I never got what theory dependent facts* are. I suppose I never did. Maybe he is more relaxed and even says: »I agree. I shall refrain from using the word ›discovery‹. I always preferred ›invention‹. You made clear to me why.« But, then, why should he not continue to use the word »discovery« applied to intra-theoretical connections, and in its perfectly intelligible, realist sense? This is not objectionable.

Isn't there quite a bit more than just intra-theoretical connections that my anti-realist may seriously use the word »discover« for? There is, but not quite as much as one might think. Remember that concepts and theories are not seriously discovered. Are experimental techniques discovered? They are rather invented, too. But what about their outcomes?

> »Newton discovered that a beam of white light sent through a prism is split up into beams of colored light.«

That this happens was a fact to discover. It would have happened before if someone had thought of making the experiment. Once the fact is discovered, it cries out for an empirically adequate theory. This may be the case with non-experimental facts, too (that there are platypuses; that, from time time, there is night on Earth). But experiments are particularly good for discovering such facts (van Fraassen calls experimental setups »engines of creation«, cf. van Fraassen (2008)). An anti-realist may, of course, say that Newton discovered that a beam of white light sent through a prism is split up into beams of colored light. This is close enough to observation.

But once more, in doing so, he would be using the word »discover« in the realist's way. For that fact is a *plain* fact, not a theory-relative one whatever that would be. He could *not* say:

> »Newton discovered that photons produce a certain quantum-electro-dynamical effect, which is observable as beams of colored light«

For he would not commit himself to the existence of photons. That is too far away from observation.

Could a realist say so? Certainly he can commit himself to the existence of photons. But would he not count the sentence as false, nay absurd, for a different reason, namely that Newton had no idea of QED? Not necessarily. Again, the sentence allows for an externalist and an internalist interpretation. Newton discovered light diffraction, but not, what it is. The case turns out to be analogous to the discovery of America with Newton in the role of Columbus, and Richard Feynman in the role of Amerigo Vespucci; and to the case of Priestley and Lavoisier. Any analysis of »discovery« in the philosophy of science should respect and get as clear as possible about this systematic ambiguity.[38]

5.7   Can the Realist Use the Verb »To Discover« as an Achievement Verb?

An objection remains to be addressed. It goes like this:

> »You presented the Einstein example as if the realist had an advantage over the anti-realist by being able to say ›Einstein discovered the fact that $E = mc^2$.‹ But then you gave a good reason for his inability to say so. For what Einstein did was analogous to what you imagined Priestley doing. Priestley did *not* discover the fact that the gas he isolated was phlogiston-reduced elementary air. According to Gilbert Ryle's useful terminology, ›to discover‹ is an achievement verb. And that gas was not phlogiston-reduced elementary air, but oxygen.«

---

38   I think that »to discover« means more than just »coming to know what one did not know before«, since the metaphor matters. However, the ambiguity sketched above might well have something to do with the knowledge component in »to discover«/»discovery« and might therefore involve tricky issues which have come to some authors' minds as issues of epistemic logic. Cf. e.g. Fitting & Mendelsohn 1998, 200f.

This is a serious objection. »To discover« is indeed an achievement verb.[39] You can only discover what is either there, or what is the case. The former case corresponds to the direct object construction. The latter case corresponds to the »that«-clause construction. »N.N. discovers that p« is veridical, just as »N.N. knows that p«. If it is true, then »p« is true. Ryle observes:[40]

> »It has long been realized that verbs like ›know‹, ›discover‹, ›solve‹, ›prove‹, ›perceive‹, [and] ›see‹ [...] are in an important way incapable of being qualified by adverbs like ›erroneously‹ and ›incorrectly‹.«

This is how the realist should reply to the objection:

> »The relevant question here is: ›Can he say so and thereby speak the *truth*?‹ And the answer is: yes. If Special Relativity is true, then the sentence ›Einstein discovered that $E = mc^2$‹ is true. That's a situation, in which he speaks the truth. So he can. Is this a situation we are in? Why not? Special Relativity is a well-confirmed theory. Is this a situation *we know* we are in? I do not think so. If I want to say something I *know*, I should say ›*I hope* Einstein that discovered that $E = mc^2$‹. That would not be saying the same thing. But all that matters here is that the verb ›to discover‹ means the same in both sentences.«

Up to »Is this a situation we are in?«, an anti-realist *like van Fraassen* would agree, and then continue: »Why care?« But then, again, he would be using »to discover« in the realist's way.

The realist's answer fits well into his overall view. Realists are humble. They pay due respect to reality.

References

Black, M. 1954. »Metaphor.« *Proceedings of the Aristotelian Society* 55: 273–294.

Carnap, R. 1931. »Überwindung der Metaphysik durch logische Analyse der Sprache.« *Erkenntnis* 2 (1931/32): 219–241.

Davidson, D. 1978. »What Metaphors Mean.« *Critical Inquiry* 5 (1): 31–47.

Deutscher, G. 2005. *The Unfolding of Language: The Evolution of Mankind's Greatest Invention*. London: Heinemann.

Einstein, A. 1994. *The Collected Papers of Albert Einstein*, vol. 5, edited by M. Klein, A.J. Kox & R. Schulmann. Princeton: Princeton University Press.

Fitting, M. & Mendelsohn, R. 1998. *First-Order Modal Logic*. Dordrecht: Kluwer.

39　Ryle [1949] 1963, 145f.
40　Ryle [1949] 1963, 146.

Fleck, L. [1935] 1980. *Entstehung und Entwicklung einer wissenschaftlichen Tatsache: Einführung in die Lehre vom Denkstil und Denkkollektiv.* Frankfurt: Suhrkamp.

Frege, G. 1892. »Über Sinn und Bedeutung.« *Zeitschrift für Philosophie und philosophische Kritik:* 25–50.

Frege, G. 1918. »Der Gedanke.« In *Beiträge zur Philosophie des deutschen Idealismus.* Band I: 1918–1919, 58–77.

Georges, K.E. 1869. *Ausführliches lateinisch-deutsches und deutsch-lateinisches Handwörterbuch aus den Quellen zusammengetragen und mit besonderer Bezugnahme auf Synonymik und Antiquitäten unter Berücksichtigung der besten Hülfsmittel.* 6th ed. Leipzig: Hahn'sche Verlagsbuchhandlung.

Gesenius, W. 1905. *Hebräisches und aramäisches Handwörterbuch über das Alte Testament.* 14th ed. Leipzig: Vogel.

Gibson, C. 2014. *Empire's Crossroads: A New History of the Caribbean.* London: Macmillan.

Glucksberg, S. & Keyzar, B. 1993. »How Metaphors Work.« In *Metaphor and Thought,* edited by A. Ortony, 401–424. Cambridge: Cambridge University Press.

Heidegger, M. [1927] 1986. *Sein und Zeit.* Tübingen: Niemeyer.

Hornfeck, S. & Ma, N. 2019. *China in kleinen Geschichten.* 3rd ed. München: dtv.

Hume, D. [1772] 1975. *Enquiries Concerning Human Understanding and Concerning the Principles of Morals,* edited by L.A. Selby-Bigge & P. Nidditch. Oxford: Oxford University Press.

Liddell, H.G. & Scott, R. 1940. *A Greek-English Lexicon* [= LSJ]. 9th ed. Oxford: Oxford University Press.

Ortony, A., ed. 1993. *Metaphor and Thought.* 2nd ed. Cambridge: Cambridge University Press.

Papineau, D., ed. 1996. *The Philosophy of Science.* Oxford: Oxford University Press.

Pfeifer, W., ed. 1995. *Etymologisches Wörterbuch des Deutschen.* München: dtv.

Quine, W.V.O. 1960. *Word and Object.* Cambridge, Mass.: MIT Press.

Ryle, G. [1949] 1963. *The Concept of Mind.* Harmondsworth: Penguin.

Searle, J.R. 1979. »Metaphor.« In *Expression and Meaning: Studies in the Theory of Speech Acts, 76–116.* Cambridge: Cambridge University Press.

Strobach, N. 2011. »Wirklichkeit im Widerstand. Vertritt Ludwik Fleck zum Realismus 1946 noch dasselbe wie 1935 und 1929?« In *Vérité, Widerstand, Development: At Work with/Arbeiten mit/Travailler avec Ludwik Fleck,* edited by R. Egloff & J. Fehr, 99–118. Collegium Helveticum Heft 12. Zürich: Collegium Helveticum.

van Fraassen, B., 1980, *The Scientific Image,* Oxford: Clarendon Press.

van Fraassen, B., 2008, *Scientific Representation,* Oxford: Clarendon Press.

Walde, A. 1906. *Lateinisches etymologisches Wörterbuch.* Heidelberg: Carl Winter.

Wittgenstein, L. [1953] 1988. *Philosophische Untersuchungen*. In *Werkausgabe in acht Bänden* [*Collected Works*], vol. 1, 224–580. Frankfurt: Suhrkamp.

Internet Resources

The following websites were helpful (last accessed May 5, 2021):

Online Etymology Dictionary: https://www.etymonline.com/, compiled by Douglas Harper. A very useful site based on reliable sources.

The multi-lingual dictionary Glosbe, maintained by a Polish team, provides a very impressive collection of lexical data for many languages: https://www.glosbe.com

Oxford English Dictionary (OED),
entry »discovery«: https://www.oed.com/view/Entry/54015?redirectedFrom=discovery#eid
entry »discover«: https://www.oed.com/view/Entry/54005#eid6715026

Making Discoveries in Physics

Kim Joris Boström

6.1 Introduction

In this article, I will elaborate on the notion of scientific discovery as introduced by Michel (2019, 2020), by distinguishing between the following types of scientific discovery:

1. abstract (or theoretical) versus concrete (or empirical) discoveries,
2. derived versus non-derived discoveries,
3. joint discoveries, and
4. null discoveries.

I will apply these concepts to selected real-world examples, and although I restrict myself to my home domain, physics, I think the proposed concepts can also be applied to other scientific domains.

6.2 The Discovery of Neptune

On the late evening of Wednesday, the 23rd September 1846, astronomer Johann Gottfried Galle (1812-1910) pointed his telescope to a rather uninteresting place in the night sky. After moving his instrument ever so slightly to and fro for a while, he spotted a pale blue dot that was not recorded in the »Berliner Akademische Sternkarte«, the stellar atlas valid at that time. His initial doubts gave way to scientific excitement, though he was still unsure about the true nature of the celestial object. Critical observation on the subsequent two nights revealed proper movement against the stellar background, and eventually Galle concluded that the celestial object was neither a star nor a comet nor a nebula, but rather a planet no one else had observed before. So, Galle is the discoverer of Neptune. Or is he?

Surprisingly, Galle refused to be acknowledged as the discoverer of Neptune and instead attributed the credits to Urbain Le Verrier, a French mathematician who had predicted the position of Neptune from observed irregularities of planet Uranus about one year earlier, in 1845. Eventually, after some back and forth, the scientific community accepted Le Verrier as the official discoverer of Neptune.[1] But how come a mathematician, not an astronomer, be

---

[1] For a very readable account on the discovery of Neptune and its aftermath, see Standage 2000.

© BRILL MENTIS, 2022 | DOI:10.30965/9783957437044_007

Figure 6.1   Planet Neptune, as seen by Voyager 2 on August 20,
               1989.

officially recognized as the discoverer of Neptune? Why Le Verrier, why not
Galle? Can a theorist, by pure reasoning alone, discover a physical object?

Moreover, Le Verrier was not the first one who predicted Neptune. John
Couch Adams (1819–1892) also noted the irregularities in the orbit of Uranus. He
calculated the position of Neptune and informed the Cambridge Observatory
in mid-September 1845, slightly earlier than Le Verrier. Adams failed, however,
to convince the director of the observatory to actually move their telescope
and look at the predicted position. He was a very polite man, and so he didn't
insist on the importance of his request, so they didn't take it too seriously. In
addition, his calculations were much less precise than those of Le Verrier, so
his instructions were rather vague.

However, not only on the theoretical side, but also on the empirical side
there were predecessors. The first observation of Neptune has been attributed
to Galileo Galilei, who apparently spotted Neptune twice, on 28 December 1612
and 27 January 1613, but then mistook it for a fixed star. So, he saw Neptune but
he didn't recognize it as a planet. This, however, is clearly essential for being
counted as the discoverer of a planet. Generally speaking, let us suppose the
following: Some person S sees some x and believes x to be an F. Let us further
suppose that x really is a G, and that being a G is essential for x, and that being

Figure 6.2   Which one of these gentlemen is the true discoverer
of Neptune?

an *F* excludes being a *G*. In a case like this, *S* would not be counted as the discoverer of *x*.

Apart from Galilei, also Jérôme Lalande in 1795 and John Herschel in 1830 observed Neptune, but both of them failed to recognize it as a planet, and so they cannot be discoverers of Neptune either.

And then there is the case of Alexis Bouvard (1767–1843). He calculated the orbit of Uranus in 1821 using Kepler's laws of orbital motion. He found, however, that Uranus did not follow the calculated orbit, but accelerated and decelerated instead. Here are four possible explanations:

A)   Bouvard's calculations are flawed.

B)   Kepler's laws are flawed.

C)   The observations are flawed.

D)   Nothing of the above, but a massive object is distorting Uranus' orbit.

Alexis Bouvard opted for explanation D and predicted that there must exist
an unknown planet. This motivated both Adams and Le Verrier to make their
own calculations and predict the orbit of the hypothesized planet. But then,
shouldn't we credit Bouvard with being Neptune's true discoverer?

There is a crucial difference between the case of Bouvard and those of
Adams and Le Verrier: Bouvard only made a *qualitative* prediction, that is,
he just concluded from his calculations that there must exist an unknown
planet, but he provided no concrete position where to observe the hypothetical planet. It seems that in order to count as a discoverer in physics, you have
to either make a crucial observation or you have to make predictions that are
precise enough to actually lead to a concrete observation. This brings us to our
first distinction.

6.3   Abstract and Concrete Discoveries

Let us make a distinction between abstract (or theoretical) and concrete (or
empirical) discoveries.

As a first approximation, a concrete (or empirical) discovery is the discovery of a concrete object like a planet, a continent, a treasure. By contrast, an
abstract (or theoretical) discovery is the discovery of an abstract object like a
mathematical truth, a physical theory, or a forgotten language. Some abstract
discoveries seem to be rather uncontroversial, such as the discovery of a mathematical relation or constant. Others, however, are more tricky. What about,
say, the fine-structure constant α that characterizes the strength of the electromagnetic interaction? It does not exist as a thing in space and time (it is a
number, after all), so shouldn't it be considered a genuinely abstract object?
Now, unlike, say, the number π, the value of the fine-structure constant is not
revealed by pure reasoning alone, but instead by controlled, repeatable physical experiments. There is something inherently concrete about this constant,
which sets it apart from π or Euler's number e, which both can be revealed to an
arbitrary degree of precision by executing mathematical algorithms. I propose
to consider the fine-structure constant a *concrete feature of the physical world*,
and hence its initial revelation a *concrete, empirical discovery*. Nowadays, the
fine-structure constant can be measured to a precision of one part in a trillion
using the Quantum Hall effect (to which we will come back later). Quantum
electrodynamics (QED) predicts a relationship between the fine-structure
constant α and the dimensionless magnetic moment g of the electron. The
measured value is consistent with the predicted relationship, which makes it
the most precise prediction a physical theory has ever produced.

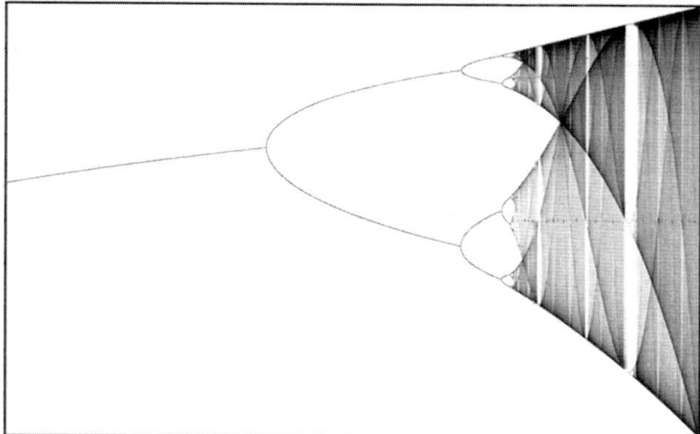

Figure 6.3   A bifurcation diagram that shows up in the
               context of deterministic chaos. The ratio between
               subsequent bifurcation points approaches the
               universal Feigenbaum constant δ.

As another example, consider the Feigenbaum constant δ. It is a universal constant that plays a central role in the context of deterministic chaos, and it was discovered in the context of describing the development of animal populations. The constant δ is a number, thus certainly an abstract object. It is the result of a theoretical reasoning about a mathematical object, the logistic equation, which happens to be a simple model for the development of animal populations under certain constraints. Unlike the fine-structure constant α, the value of the Feigenbaum constant can be revealed by mathematical reasoning alone, just like π or e, namely by calculating the limiting ratio of subsequent bifurcation intervals of the logistic equation. Hence, the Feigenbaum constant is an abstract object that is motivated, though not determined, by concrete empirical observation. It is a feature of the physical world only in so far as it appears in a mathematical description of the physical world, in which regard it is similar to π that appears in many physical laws describing nature. Hence, the discovery of the Feigenbaum constant is an abstract, theoretical discovery.

Now, what about the discovery of a biological species (cf. Michel 2019)? While individual animals exist in space and time, a biological species in itself does not exist in space and time, it is a mere abstraction. Despite being an abstraction, however, the concept of a biological species is entirely grounded in observing nature. Imagine biologist Charles (cf. Michel 2019, 428) awakes from his dreams about a cat having a fish tail, and he then goes and officially announces the discovery of a new species named »catfish«. He would certainly reap the derision of his contemporaries, and not just because the name »catfish« was already reserved. To discover a biological species requires concrete, empirical observation, it cannot be done in theory (or in a dream) alone. Hence, the discovery of a biological species is a concrete discovery, although the species itself is abstract.

Next, let us take a look at the discovery of a *physical effect*, such as the photo effect or the Quantum Hall effect. An effect is not a concrete object in time and space, but still one may point to a burst of smoke, a flashing light, a detector display, or the dip in a plotted time series of measurement values, and shout »See, the effect, there it is!«. Similar to a biological species, a physical effect is in itself an abstract object, but is inherently grounded in empirical observation. It is impossible to discover a physical effect without doing some concrete, objective research that can be replicated by other researchers. This does not exclude the possibility to theoretically *predict* a physical effect (and we will come back to this later on), but the prediction itself is not sufficient to constitute a discovery. The discovery of a physical effect must be supported by empirical observation, and hence it is a concrete discovery.

Lastly, how should we categorize the discovery of a *physical theory*? A physical theory is certainly an abstract object, but it is also grounded in empirical observation, similar to a physical constant. If a physical theory, which is basically a set of physical laws, does not yield testable predictions that match the observations, it is not a useful theory. If we modify, say, Newton's second law $F = m \cdot a$ into $F = m \cdot a^2$, because we think it looks prettier, as it now resembles Einstein's famous equation, then we obtain a theory that is pretty useless, because it is immediately falsified by observation. A physical theory must be *compatible with* empirical observations, but this relation to nature is much weaker than the relation of a physical constant to nature. The value of a physical constant can be defined by a specific class of concrete, repeatable experiments. Physical laws, on the other hand, describe abstract relations rather than concrete observations. Newton's second law amounts to saying »If I were to accelerate an object of mass m by an amount a, then I would need a force of strength F. And vice versa, if I put a force of strength F against a body of mass m, then the body will accelerate by the amount a.« No matter what the individual values for F, m and a actually are. Physical laws are never concrete statements, such as the statement »The fine-structure constant has a value of $\alpha = 0.007297352 \ldots$«; this statement not a physical law. Most physical laws come in the form of differential equations. They only describe a concrete observation if other concrete information is provided, such as the initial conditions and the parameter values. Without this concrete information, physical laws do not describe anything in nature; they are just abstract, mathematical equations. Therefore, I suggest to count the discovery of a physical theory, equation, or law, as an abstract discovery.

In the view of the above considerations, the discovery of Neptune is a concrete one. For one thing, because the discovery was based on empirical observations of a concrete object in the first place, namely the observed irregular motion of Uranus. Unlike the case of Feigenbaum's constant, these empirical observations did not just motivate Le Verrier's calculations. Rather, his calculations are fully dependent on those observations. The distorted trajectory of Uranus, and hence Uranus itself, is an irreducible empirical element of the calculations of the position of the hypothesized planet Neptune. If the observations had been flawed, or if they had been reported falsely, the predicted position of Neptune had turned out differently and Neptune would not have been discovered. Second, the object of discovery, planet Neptune, is a concrete object. This all makes the discovery of Neptune a concrete discovery.

6.4   Joint Discoveries

So, did Le Verrier, the mathematician, made a concrete, empirical discovery, when he predicted the position of Neptune? Did he discover Neptune »with the point of his pen«, as François Arago, the director of the Paris observatory, famously put it? Was the discovery completed when Le Verrier published his calculations? What if Galle had not pointed his telescope to the sky to actually observe the planet at the predicted position? I argue that without actual observation of the hypothesized planet at the predicted position, Le Verrier would not have been regarded as a discoverer at all. He would not even have made an abstract, theoretical discovery, for his calculations would bear no significance whatsoever, if they would not refer to a materially existing object: planet Neptune. There is no practical value in the calculation of the orbit of a non-existing object.

Vice versa, without Le Verrier's calculations, Galle would not have pointed his telescope to the correct position in the sky to find Neptune. Sure, if he had spotted Neptune by chance, as Galileo did, and if he had then drawn the correct conclusions and identified the observed object as an unknown planet, then, of course, he would have counted as the true discoverer of Neptune. However, as a matter of actual fact, he was motivated by Le Verrier's calculations, and *only* by Le Verrier's calculations, to point his telescope to a particular spot in the sky and find Neptune. Concluding, only together in tandem, the theoretical prediction together with the empirical observation make the concrete discovery.

But what about the other precursors of the discovery? Was not Le Verrier motivated by Bouvard to do his calculations? And was not Bouvard led to his conclusions by observations of other astronomers? I think the case becomes clearer when one considers which of these actions were not just necessary but *essential* for the discovery of Neptune. Certainly, there is a causal chain running through several astronomical observations and theoretical considerations by numerous people, leading to the prediction of Neptune's position by Le Verrier and finally to the empirical observation by Galle. If any event within the chain had not occurred, the final observation might not have taken place or might not have turned out successful. But the events do not equally contribute to the final successful observation. The merits of a discovery should neither always be attributed to the last member of the chain only, nor to the entire chain of people whose actions have in some way or the other contributed to the successful discovery. If there are several people whose contributions are essential and more or less equally important for the success of the discovery, then the discovery should be considered a *joint discovery*[2], and its merits should be shared among its contributors. The importance of a contribution thereby depends on the *amount of necessary effort*[3] that has been put into the process of discovery to make it successful.

Now, in the light of these considerations, it appears more plausible why Galle had refused to be called the discoverer of Neptune, and why the scientific community and the general public accepted his refusal so willingly. They all felt that Galle's contribution to the discovery, although representing a necessary completion to the entire *discovery process* (Michel 2019, Michel 2020), was simply not as important as that of Le Verrier. Le Verrier did all the work and Galle simply had to point his telescope to the position Le Verrier had calculated. Pointing the telescope to a predetermined position is a fairly easy task for a professional astronomer, and something he would do on a regular basis. This piece of work, although being indispensable for the discovery to be completed, was simply not important enough to grant him all the merits of the discovery of Neptune.

In 1928, in an attempt to make the central equation of quantum mechanics, the Schrödinger equation, compatible to Einstein's theory of special relativity, young physicist Paul Dirac (1902–1984) arrived at an elegant, but somewhat peculiar equation that was soon named after him, the *Dirac equation*:

$$i h \gamma^\mu \partial_\mu \psi - mc\psi = 0.$$

---

2   The concept of a joint discovery has been introduced by Michel (2019, 430) who also calls it a *joint effort*. My proposal builds upon the original concept.

3   Unnecessary effort doesn't count. John cannot claim that his contribution to the discovery of an Egyptian tomb is more important than Joe's only because John excavated the tomb by hand while Joe used machines.

The equation was indeed relativistic, so it complied with Einstein's theory, and it transformed into the Pauli equation in the non-relativistic limit, $c \to \infty$. This was very exciting, because the Pauli equation had been discovered by Wolfgang Pauli (1900–1958) just one year before, and it accurately described an electron with a novel property called »spin« (Pauli 1927). Since Dirac's equation reproduced Pauli's equation in the non-relativistic limit, it offered an *explanation* for the novel and rather abstract spin property. However, Dirac's new equation had an additional class of solutions: It not just described an ordinary electron having the correct mass, negative charge and spin; it also described a particle that has all those electron properties except having negative energy. This seemed to be fatal, as it was firmly assumed that the energy of a particle must always be positive or else matter would become unstable and decay, releasing an infinite amount of energy.

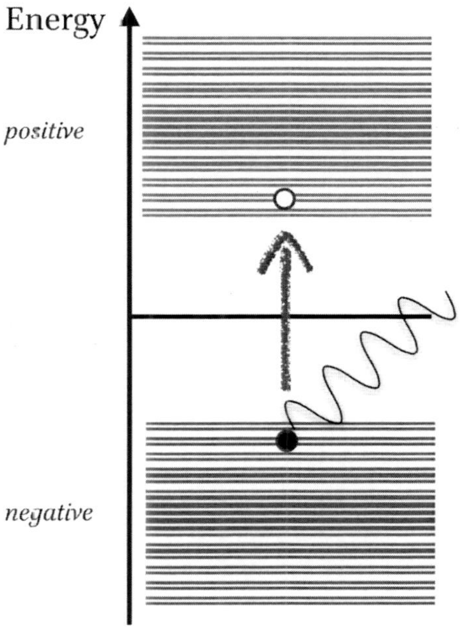

Figure 6.4
Creation of antimatter (more precisely, of a particle-antiparticle pair) in the Dirac sea interpretation. A high-energy photon (wiggly line) hits an invisible, negative-energy electron in the infinite Dirac sea (lower lines), which absorbs the energy and jumps onto a positive energy level (upper lines), thereby becoming an ordinary, visible electron o. The remaining hole in the Dirac sea (black dot) acts like a positive-energy, thus visible anti-electron of positive charge: the positron.

As a way out of this dilemma, Dirac proposed the following solution: All states of negative energy are occupied by an infinite sea (later called the *Dirac sea*) of particles. Due to *Pauli's exclusion principle*, positive-energy electrons cannot make a transition to these already occupied negative-energy states, so that matter remains stable and, moreover, we would not notice these particles as they would remain inert. Now, if a photon of sufficient energy is absorbed by a negative-energy electron, the electron jumps to a state of positive energy, thereby becoming an ordinary electron and leaving a hole in the negative-energy sea. This hole would effectively possess positive energy and, hence, would appear as an ordinary particle. Being a hole in a sea of negatively-charged electrons, however, this »anti-electron« would possess positive charge. Dirac at first thought he had found the theoretical foundation of the proton.

There is an intriguing analogy to the electronic band structure in a semi-conductor: The lower-energy bands are occupied with electrons that remain inert, and only those electrons that absorb enough energy from the environment and jump to a higher-energy level can freely move, become electrically active and contribute to the conductivity of the semiconductor. But not just that, also the remaining hole in the lower-energy band becomes electrically active, can freely move and contributes to the conductivity, thus becoming a *quasi-particle*.

However, Dirac's proposal was met with criticism:
1. The proposal implied an infinite negative charge everywhere in space, which was not observed.
2. Electrons and protons should have the same mass, which was not observed.
3. Electrons and protons should annihilate each other, which was not observed.

Dirac then modified his proposal to address at least the two last issues: The hole in the Dirac sea does not correspond to a proton but rather to a hypothetical, positively charged anti-electron, the *positron*. Still, the problem of the infinite negative charge of the universe remains, and also the whole hole theory does not work for bosons. So, nobody was really convinced.

Then, in 1932, Carl David Anderson (1905–1991) actually found the positron. He investigated cosmic rays using cloud chambers and found traces that he interpreted as particles having the same mass as electrons but opposite charge. Immediately, he identified them as the positrons predicted by Dirac, further confirmed his results by actively creating electron-positron pairs with gamma rays and published his observations. Dirac's crazy idea suddenly seemed to be confirmed, very much to the surprise of the scientific community, in particular Wolfgang Pauli and Niels Bohr, the leading quantum physicists at the time, who had strongly argued against Dirac's interpretation before (Crețu 2020; Dittrich 2015). Later on, Anderson admitted that he was inspired by the work of his Caltech classmate Chung-Yao Chao (1902–1998), who, in 1929, studied the scattering of gamma rays in lead. Chao had obtained anomalous results indicating particles behaving like electrons, but with a positive charge. He had published his observations, but not offered an interpretation. Similarly, in 1929, Soviet physicist Dmitri Skobeltsyn had detected particles in his cloud chamber, which acted like electrons but had opposite charge. His observation remained widely unrecognized, as he did not offer an interpretation (and probably also because he was on the »wrong« side of the iron curtain). Also, Irène and Pierre Curie had observed positrons, but they dismissed them as protons. Yet another observation of positrons had been made by Patrick Blackett and Giuseppe Occhialini earlier in 1932. However, they delayed the publication of their results to obtain more solid evidence. Their laudable scientific thoroughness would turn out rather unhappy, as in the same year Anderson published his observations and won the physics Nobel Prize of 1936 as the discoverer of the positron.

So, who really is the discoverer of the positron? First of all, should not Paul Dirac, analogue to Le Verrier, be called the discoverer of antimatter?

On the pro side of the argument we have:

1. Dirac derived a new equation and found an anomalous solution.
2. He interpreted the anomalous solution as describing a novel particle, which was then experimentally observed.
3. Another prediction of his was that the new particle may also annihilate with an electron into pure energy, which had also been confirmed later on.

However, on the contra side we have:

1. Dirac initially misinterpreted the new particle as a proton.
2. His »hole theory« is not very elegant and has unresolved issues.
3. He did not convince anybody to conduct experiments to test his prediction.
4. His interpretation became obsolete by the advent of quantum field theory.

As for the latter point, a much more elegant and consistent interpretation of the Dirac equation was found in the context of quantum field theory. In this context, the wave function does not describe a particle but rather a *quantum field* whose excitations are particles. Negative-energy solutions would correspond to the creation of positrons. In this interpretation, no infinite sea of negative-energy particles is necessary.

Dirac won the Nobel Prize anyway, namely in 1933 for »for the discovery of new productive forms of atomic theory« (cf. Larsson and Balatsky 2019). He is still acknowledged to have predicted the positron, so for him the story has a happy ending. But what about the Dirac equation itself? Is it not an abstract, theoretical discovery? Against the background of the distinction between concrete and abstract discoveries introduced above, the Dirac equation in itself represents an abstract discovery. For the first time, someone proposed a first-order differential equation that complied to the theory of special relativity and that transformed into the Schrödinger equation in the non-relativistic limit. Even if empirical evidence had shown that the Dirac equation does not describe anything existing in observable nature, the equation would still have its value as a possible, theoretical solution to a problem in theoretical physics. This brings us to the next case.

6.5   Null Discoveries

At the end of the 19th century, light was firmly believed to be a wave. In 1864, James Clerk Maxwell (1831–1879) presented his groundbreaking equations (the *Maxwell equations*) to the Royal Society, and so described electric and magnetic phenomena by one unified theory called *electromagnetism* (Maxwell 1865). Although not necessary for the interpretation of his equations, Maxwell's mindset was built upon the concept of a hypothetical material medium, the luminiferous ("light-bearing") ether, that permeates the entire universe, and the mechanical undulations of which represent electromagnetic radiation including light, gamma rays and radio waves. In an earlier paper, Maxwell drew a rigid grid of hexagonal vortices that rotate against each other, thereby pushing and pulling charged particles around, which represents electric current, and whose oscillatory rotations generate a spatially propagating wave that represents electromagnetic radiation (Fig. 5, background, from Maxwell 1861 Fig. 2). The propagation speed would reach a maximal, constant value c in the absence of any other material than the ether itself. The success of Maxwell's equations in explaining all electromagnetic phenomena, qualitatively as well as quantitatively, was so convincing that no one any longer doubted the existence of the luminiferous ether. It seemed that the century-old struggle between proponents of the wave theory of light, prominently led by Christiaan Huygens (1629–1695), and the corpuscular theory of light, prominently led by Isaac Newton (1642–1726), was finally settled. The wave proponents seemed to have won a glorious victory that was crowned by the unification of two hitherto distinct phenomena, electric and magnetic forces. Researchers no longer debated the existence of the ether, but just the exact properties of this substance. Based on hydrodynamical investigations of Hermann von Helmholtz (1821–1894), who demonstrated that vortex rings are indestructible in an ideal, frictionless fluid (v. Helmholtz 1858), Lord Kelvin (1824–1907, also known as Sir William Thomson) proposed that the atoms of matter were just such vortices within the ether (Thomson 1869), William Mitchinson Hicks (1850–1934) refined Kelvin's theory into a »vortex sponge theory« where the atoms were thought to be constituted of larger sponge structures instead of single vortices, and so forth (cf. v.d. Laan 2013). The density and elasticity of the ether became a matter of controversy, since in order to propagate waves at the speed of light, the ether would have to be very rigid, and at the same time it had to be fluid or gaseous to allow material objects to move across without resistance.

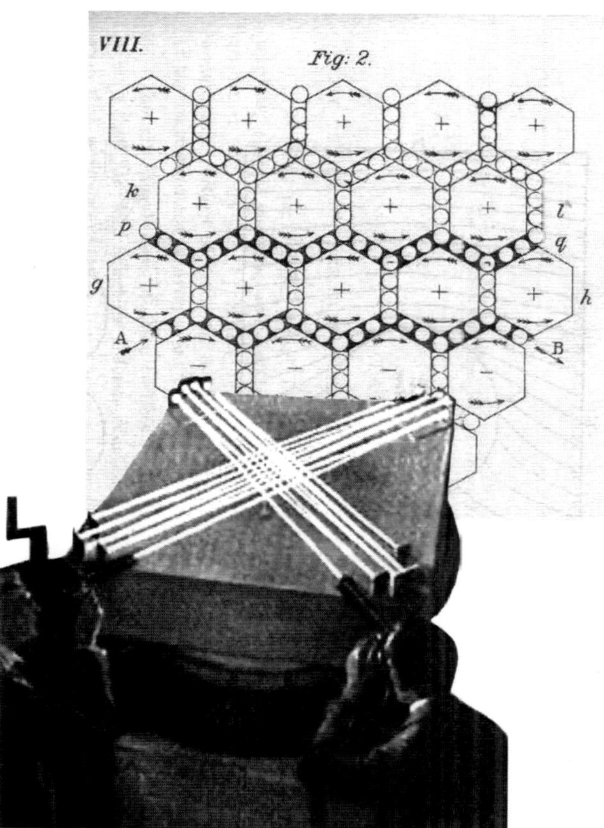

Figure 6.5   Foreground: Michelson and Morley at their self-developed interferometer. Background: The ether as conceptualized by Maxwell and contemporary scientists, a »sponge« of »molecular vortices«, whose oscillatory rotation propagates through space and constitutes electromagnetic radiation.

And then the Michelson-Morley experiment happened. Researchers were already agreeing that if the ether exists then the earth was moving through it, and this relative motion, the »ether wind«, should actually be measurable. In 1887, Albert Michelson and Edward Morley conducted an experiment using a self-developed, highly sensitive interferometer to measure the ether wind. They found: nothing.

Now, does the null result of Michelson and Morley count as a discovery? They found nothing, no ether wind, no expected effect of the relative motion of the earth through the ether – so can this nothing be the object of a discovery? Wouldn't that be contradictory or at least meaningless? I propose that it is perfectly reasonable to count a case like this as a discovery. The apparent contradiction stems from an overly reduced focus on the object of discovery. If there is no object, so it seems, there can be no discovery. However, also the finding *that something is the case* may count as a discovery. Consequently, also the finding *that something is not the case* may count as a discovery. In analogy to the notion of an experimental null result, I propose to call the discovery that something is not the case a *null discovery*.

So, does the Michelson-Morley experiment count as a null discovery then? If so, it would be the concrete discovery that the ether, as the hypothetical substance that permeates the entire universe, whose undulations represent electromagnetic radiations, etc., does not exist. However, there are several arguments to *not* count the Michelson-Morley experiment as a (null) discovery:

1. Michelson and Morley themselves did not draw the conclusion that the ether does not exist. Rather, they tried to explain their result by assuming that the earth drags the ether with it, so that no measurable relative motion exists.
2. The scientific community also did not draw the conclusion of the non-existence of the ether.
3. Hendrik Lorentz (1853–1928) explained the null result with a hypothesized contraction of the interferometer arms in the direction of movement, in consistency with the existence of the ether.

It was not until 1905, when Albert Einstein (1879–1955) laid the foundation of the theory of special relativity (SRT) in his landmark paper (Einstein 1905b), that the correct conclusions about the non-existence of the ether were convincingly drawn. Einstein not only showed that the conception of the ether was obsolete, he also presented the outlines of an entirely new theory that would revolutionize not just theoretical and experimental physics but the way we perceive the world. So, interestingly, only Einstein's abstract discovery of a new theory completed the concrete discovery of the non-existence of the ether.

Today, it is generally accepted that the ether does not exist, at least not in the way its original defenders believed it to exist. Light can propagate through empty space, which is not a medium in the material sense, although it has physical properties, such as curvature and expansion. Hence, there is a way in which one may still believe in an existing ether, if one insists in doing so, although not in the way originally intended by its proponents. As Einstein himself put it (Einstein 2007):

> Recapitulating, we may say that according to the general theory of relativity space is endowed with physical qualit_es; in this sense, therefore, there exists an ether. According to the general theory of relativity space without ether is unthinkable; for in such space there not only would be no propagation of light, but also no possibility of existence for standards of space and time (measuring-rods and clocks), nor therefore any space-time intervals in the physical sense. But this ether may not be thought of as endowed with the quality characteristic of ponderable media, as consisting of parts which may be tracked through time. The idea of motion may not be applied to it.

In sum, the non-existence of the ether (in the original sense) is a concrete, joint discovery by Michelson and Morley, who provided the empirical investigations, and by Einstein, who made a theoretical contribution by drawing the correct consequences. There is a striking similarity to the discovery of Neptune, where Le Verrier made a theoretical prediction and Galle completed the discovery with a concrete observation. In the Michelson-Morley case, however, the temporal ordering is reversed: While Le Verrier made the theoretical predictions before Galle confirmed them by empirical observation, in the case of the ether, Michelson and Morley made the empirical observations and then Einstein drew the correct conclusions.

6.6   Derived and Non-derived Discoveries

It wasn't, however, Michelson and Morley's null result that triggered Einstein's development of the Theory of Special Relativity.[4] Rather, he was deeply worried about a theoretical loophole in the foundations of physics: Classical Newtonian Mechanics is Galilei-invariant, that is, Newton's laws look the same in all reference frames moving at constant speed. In contrast, the laws of electrodynamics, the Maxwell equations, are not Galilei-invariant, that is, they do not look the same in all moving reference frames. However, the observed

---

4   In his 1905 paper, Einstein refers to »unsuccess_ul attempts to establish a movement of the earth relative to the ›light medium‹ « (Einstein 1905b, 891 my translation). It remains controversial whether he referred to the Michelson-Morley experiment or (also) to experiments by Arago, Fizeau, Hoek, and many others, all of which failed to yield conclusive support for the existence of the ether (cf. Wikipedia contributors 2020).

effects are nonetheless always identical, so the laws *should* look the same, shouldn't they?

Einstein took these considerations so seriously that he postulated the *principle of special relativity*: The laws of physics must look the same in all inertial frames (these are frames that move with constant speed relative to each other). Secondly, he took another fact very seriously, namely that the Maxwell equations contain one parameter that retains its value in all inertial frames: the vacuum speed of light c. Thus, he raised this mathematical circumstance to a postulate: the vacuum speed of light must be the same in all inertial frames. Now, these two postulates are really equivalent to the following one postulate that *the laws of physics must be invariant under Lorentz transformations.* Indeed, it's the Lorentz transformations that leave the Maxwell equations invariant.[5] As Hermann Minkowski (1864–1909) realized, Einstein's postulates lead to the conclusion that time and space are not separate from each other but form the unified, four-dimensional *spacetime*. The occurrence of events within spacetime are not absolute but rather relative to the state of motion of the observer. For observers that are in motion relative to one another, the same event occurs at distinct points in space and time, spatial distances and temporal duration may differ, there is no absolute »now«, and the order of (causally disconnected, »space-like«) events may even reverse. So, in this sense each observer exists in their own spacetime.

These mind-blowing conceptions were even further extended by Einstein's discovery of the general theory of relativity (GRT). Basically, the general theory replaces the special one by postulating that *the laws of physics must look the same in all freely falling reference frames.* In the limit of zero gravity, free fall becomes constant motion, so that freely falling reference frames become inertial frames, and thus the general theory of relativity reduces to the special theory. The consequences of this straightforward generalization of the principle of relativity are enormous. Spacetime now behaves as a material rubber cloth rather than an abstract reference frame: it is compressed, stretched and curved by the presence of mass and energy, and its curvature acts back on the movement of massive objects and even on the propagation of light.

Einstein's contributions to theoretical physics cannot be exaggerated. They are revolutionary in every respect. However, it has been argued that Henri Poincaré (1854–1912) was actually the first one to discover special relativity. Indeed, Poincaré discovered the Lorentz transformations, and showed that

---

5 The largest group of transformations that leave the Maxwell equations invariant is the *conformal group* of which the Lorentz group forms a subgroup.

they leave the Maxwell equations invariant (Poincaré 1905). He reasoned about the impossibility of measuring the ether wind, although he didn't question the ether (Poincaré 1904). He formulated his own version of the principle of relativity, but he only thought of it as describing what we can measure, not what there is (Poincaré 1904). He came up with the equation $E = mc^2$, but he only applied it to the electromagnetic field, which in his view only effectively can be conceived of as a »fictional fluid« possessing a mass density, and he rejected it to be conceived of as a »real fluid« (Poincaré 1900). In short, he saw the trees but not the wood. Einstein, on the other hand, did see the trees as well as the wood. In addition, he saw the entire landscape with all the trees, bushes, birds, bears, and bees. Thus, Poincaré is definitely an important and relevant precursor of the theory of special relativity, and he deserves the merit of several theoretical discoveries that form its basis. But as for the final theory of special relativity, Einstein was the one who connected the dots and drew the full picture.

As for the general theory of relativity, there is general consensus that Einstein deserves the full credits to have discovered the theory. However, there is some controversy about the role of renowned physicist David Hilbert (1862–1943, cf. Thorne 1994; Corry 1997; Todorov 2005; Veisdal 2019). As a matter of fact, Hilbert was also working on a relativistic theory of gravitation, and he published the field equation five days before Einstein. It is, however, unclear if Hilbert's version of the field equation originally deviated from Einstein's final version, and if the changes made by Hilbert to his equation were inspired by Einstein or, on the contrary, if Einstein »nostrified« his equation from Hilbert. Corry (1997) pointed out that in the first draft of Hilbert's publication, which was sent to him for proofreading, he added to the field equation the hand-written remark »as first introduced by Einstein«. Also, Hilbert always fully credited Einstein as the originator of the theory, and there was no priority dispute between the two men afterwards. In the words of Kip Thorne (1994):

> Quite naturally, and in accord with Hilbert's view of things, the resulting law of warpage was quickly given the name the Einstein field equation rather than being named after Hilbert. Hilbert had carried out the last few mathematical steps to its discovery independently and almost simultaneously with Einstein, but Einstein was responsible for essentially everything that preceded those steps.

So let us take a glimpse at that ominous field equation:

$$R_{\mu\nu} - \frac{1}{2} R g_{\mu\nu} = \kappa T_{\mu\nu}.$$

This elegant equation uses very condensed mathematical notation and actually represents a system of ten coupled non-linear partial differential equations. These equations can be applied to describe the movement of planets and stars, of galaxies and nebula, to describe the stretching and curving of spacetime by energy and mass, and they are very difficult to solve. One of the solutions is the following. If a massive object is compressed below a certain size, it turns into a black hole, a spacetime singularity from where nothing, not even light, can escape. The theoretical possibility of black holes was discovered in 1915 by Karl Schwarzschild (1873–1916). After him, the radius of a spherical body below which it turns into a black hole was named the *Schwarzschild radius*, and it is directly related to the mass of the body. For example, the Schwarzschild radius of the sun is 3 km, that of the earth is 9 mm, and that of Messier 87, the supermassive black hole which had famously been photographed in 2019 is 20,000,000,000 km (Fig. 6).

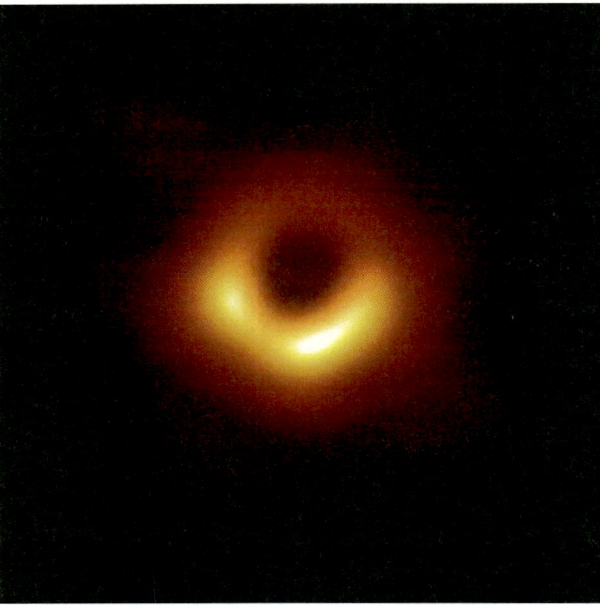

Figure 6.6   The famous photo of the black hole Messier 87
                (abbreviated M87) in the center of a distant galaxy
                of the same name. (To distinguish them, often an
                asterisk is put behind the name of the black hole, as
                in M87*.)

The black hole discovered by Schwarzschild is not a real black hole. It is a mathematical solution to Einstein's field equations, which form the mathematical basis of GRT, so Schwarzschild's discovery was a theoretical one. Moreover, most scientists at the time thought that black holes did not exist as real objects. They considered Schwarzschild's solution as an abstract mathematical artifact without physical relevance. Moreover, it seemed impossible that black holes could somehow be formed by any real-world physical process. And finally, black holes are, well, black, so they cannot be directly observed anyway, which seemed to make them even more phantasmagoric. But then in the late 1960s Roger Penrose (*1931) and Stephen Hawking (1942–2018) showed that black hole solutions are mathematically perfectly sound, and that black holes are not just possible but rather the inevitable fate of any star about 5 times more massive than the sun, of which there are plenty in the observable universe (cf. Ford 2003; Ellis 2014). Over the decades, the sentiment changed from pure denial to firm belief. More and more empirical evidence piled up to support not just the possibility but the actual existence of black holes. Astronomers observed the orbital movement of stars around an invisible object in the center of the Milky Way, incredibly powerful gamma bursts and lightning fast plasma jets expelled from the active centers of galaxies, and, in 2015, gravitational waves that behave exactly as if they were created by the merging of two supermassive black holes (Abbott et al. 2016). But it wasn't until 2019 that the Event Horizon Telescope, a virtual telescope that results from combining the observational data from 8 radio-telescopes around the earth, finally provided the first direct observation of a black hole (Akiyama et al. 2019). The journal *Science* declared this the *Scientific Breakthrough of the Year 2019*.[6]

So, does this finding represent the completion of the concrete discovery of a black hole as theoretically predicted by Schwarzschild? The Schwarzschild case is actually very similar to the case of Bouvard, who predicted the existence of Neptune from irregularities of the orbit of Uranus. In contrast to Le Verrier, Bouvard did not make a quantitative prediction that would directly lead to the discovery of Neptune. Similarly, Schwarzschild predicted the existence and some essential properties of black holes from Einstein's equations, but he did not provide any details of how and where to actually find them. He was not even sure whether the solutions he discovered actually corresponded to physically existing objects. Thus, although Schwarzschild's black hole solution may be a remarkable theoretical discovery, it is not part of the concrete discovery of Messier 87 or any other physically existing black hole. But is the black hole

---

6   See https://vis.sciencemag.org/breakthrough2019/finalists/#Darkness-made-visible.

solution actually a discovery at all? Considered closely, Schwarzschild found a solution to equations that Einstein had already put up. The solutions were implicit in the equations, but only nobody noticed them before. This brings us another distinction I would like to draw, namely that between *derived and non-derived discoveries*. A derived discovery is implicit in another discovery; it merely reveals what is already there, although this revelation may appear surprising enough to be perceived as a proper discovery on its own. Consider Newton's famous three laws. They are certainly a non-derived theoretical discovery: No-one has ever come up with three fundamental laws that explain such a broad range of natural phenomena. Any mechanical phenomenon and any phenomenon involving gravity, be it the dancing of dust particles in the air or that of planets and stars in space, can be explained to an overwhelming degree of precision from only these three laws. Together with Maxwell's equations, these theories formed the basis of the physical description of nature at the end of the 19th century. Nothing seemed to evade the explanatory power of these classical theories, which had Philipp von Jolly (1809–1884), a physics professor of Max Planck, led to famously discourage his student from going deeper into physics. Planck reports the following (Planck 1943, 1:102, translation by Jan Michel and me):

> [My honorable teacher Philipp von Jolly] portrayed to me physics as a highly developed, almost completed science, which, after having been crowned by the discovery of the principle of energy conservation, would probably soon have arrived at its final stable form. Admittedly, in one corner or another there might still be a speck of dust or a tiny bubble to examine and to classify, but the system as a whole would stand fairly safe, and theoretical physics would clearly be approaching the level of perfection that, for instance, geometry had already possessed for centuries.

And then, in 1900, while trying to tackle the problem of black body radiation, Planck discovered that in his calculations the radiation must be absorbed and emitted in small packages of energy which he called »quanta«. This very clever mathematical trick, which helped to solve the problem of black body radiation, eventually led to the development of Quantum Theory, the biggest physical revolution apart from the theory of relativity (Planck 1901). Planck's discovery was a non-derived one. His quanta were not in any way contained within the framework of the classical theories, they represented an absolutely novel element in the physical description of nature. However, Planck was very reluctant to grant his discovery any physical existence at first. Also Einstein, who used Planck's quanta to explain the photo effect in the same year, was careful and called his analysis a »heuristic viewpoint« (Einstein 1905a). But the devil was out of the box, and soon the scientific community hotly debated not just the existence, but the exact properties and behavior of »photons« and other »quantized« particles.

Fast forward to 1961: American meteorologist Edward Lorenz investigates weather prediction models using his trusty analog computer. He comes up with a nice and simple model, and finds that it is impossible to reproduce the same weather situation beyond a certain simulated period of time, despite entering the same initial conditions into the computer (Lorenz 1963). This physical system, so it seemed, was unpredictable. The basic observation that there are unpredictable systems, however, was not entirely new. There had been considerations that certain physical systems behave unpredictably, especially by Henri Poincaré as he investigated the three-body problem (Poincaré 1890). But still, it was widely believed that nature is overall well-behaved and, in principle, predictable. Some twenty years after Lorenz‹ befuddling observation, Benôit Mandelbrot (1924–2010) found that unpredictable behavior is not the exception but rather the norm, and he coined the term »fractal« to denote those mathematical beasts that govern the unpredictable behavior (Mandelbrot 1982). Finally, the scientific community began to accept the physical relevance of *Deterministic Chaos* – the apparently erratic, unpredictable behavior of non-linear systems that still obey the deterministic laws of classical physics.

The discovery of Deterministic Chaos by Poincaré, Lorenz and Mandelbrot is a derived theoretical discovery. Deterministic Chaos lingers within a certain very common, even generic class of differential equations, which all derive from the three laws of Newton. Chaotic systems, not just complex ones like the weather but also simple ones like the double pendulum, are fully described by classical Newtonian mechanics, and still they behave in an unpredictable manner. This means that deterministic chaos has been there all along, it was only hidden within the complex consequences of Newton's deceivingly simple equations.

As a last example, consider the *Quantum Hall effect*, which has been discovered in 1980 by Klaus von Klitzing (*1943), and for which he received the physics Nobel Prize in 1985. The effect shows up as follows. In the ordinary Hall effect, electric current flows through a thin metal plate, a magnetic field is applied to the plate, and the electric flow is deviated by the magnetic field. Now, for very low temperature and strong magnetic field, the Hall resistance ρ_{xy}, which is the transverse to the electric current, undergoes discrete jumps as the magnetic field increases. These discrete jumps are integer multiples of the *von Klitzing constant*

$$R_K = \frac{h}{e^2} = \frac{1}{2c\epsilon_0\alpha}.$$

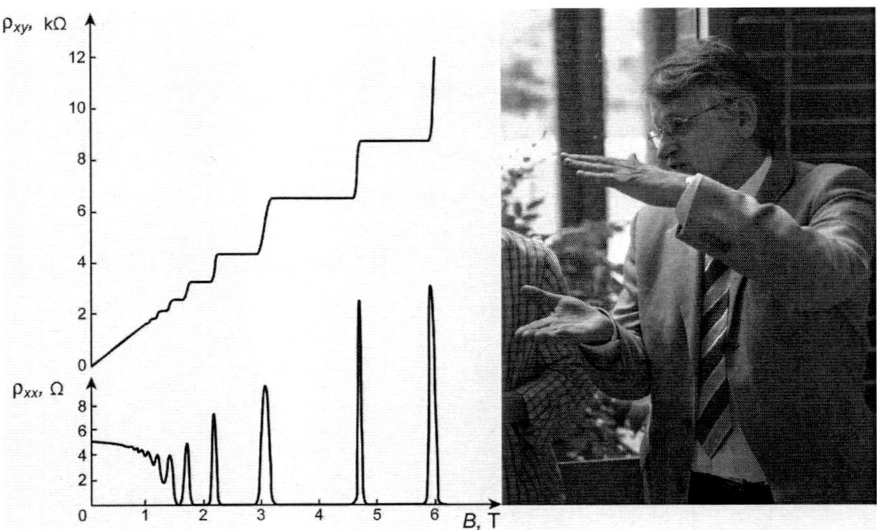

Figure 6.7   The quantum Hall effect and its discoverer, Klaus von Klitzing. The Hall resistance ρxy (upper y-axis) undergoes discrete jumps as the magnetic field strength B (x-axis) is increased.

The von Klitzing constant is a universal constant that can exactly be reproduced. It is now used to determine the value of natural constants h, e, and α.
So, it is a concrete feature of the physical nature, and its discovery is thus a
concrete discovery. But: The quantum Hall effect is entirely implicit in the laws
of quantum mechanics. Just like the Schwarzschild solution, the Quantum
Hall effect lingered within the already known laws of Quantum Mechanics,
waiting to be revealed by a smart enough scientist. Still, the theoretical work
of unveiling the effect by theoretical considerations, together with the actual
experimental work leading to its empirical observation, constitutes a great
achievement and a big enough novelty to be considered a discovery in its own
right, albeit a derived one.

6.7   Conclusions

In this contribution, I have analyzed several discoveries in physics and have
proposed some new terminology to distinguish and identify certain types of
discovery. These types and in particular the proposed definitions are certainly
not set in stone. They can and should be improved in actual practice. I hope
that they will sharpen the sensitivity towards scientific discoveries and help to
analyze real-world cases. I would like to particularly emphasize the importance

of accepting certain discoveries as joint discoveries. People tend to think in an exclusive way, asking »who is the discoverer now?«, when actually there is reason to accept more than one person as the discoverer of the same thing. Also, the null discovery deserves more attention and appreciation. To find that something is not the case can be just as important as the positive case. In the scientific literature, we have the general problem that null results (which are, in many cases, null discoveries) are inherently hard to publish, because they do not seem sensational enough. This attitude should change.

Acknowledgements

I would like to thank Jan Michel and Gernot Münster for exciting discussions and extensive feedback to my manuscript draft.

List of Figures

Figure 6.1
Neptune, photographed by Voyager 2 on August 20, 1989.
Credits: NASA/JPL/Voyager-ISS/Justin Cowart
License: Creative Commons Attribution 2.0 Generic
Source: https://commons.wikimedia.org/wiki/File:Neptune_-_Voyager_2_
(29347980845)_flatten_crop.jpg

Figure 6.2
contains four panels:
– Johann Gottfried Galle
Credits: unknown
License: Public Domain
Source: https://commons.wikimedia.org/wiki/File:Johann-Gottfried-Galle.jpg
– Urbain Le Verrier
Credits: Magnus Manske on en.wikipedia
License: Public Domain
Source: https://commons.wikimedia.org/wiki/File:Urbain_Le_Verrier.jpg
– John Couch Adams
Credits: unknown
License: Public Domain
Source: https://commons.wikimedia.org/wiki/File:John_Couch_Adams.jpg

– Alexis Bouvard
Credits: Jules Boilly
License: Public Domain
Source: https://commons.wikimedia.org/wiki/File:Alexis_Bouvard.jpg

Figure 6.3
Bifurcation diagram
Credits: Kim Boström
License: Unrestricted redistribution, commercial use, and modification

Figure 6.4
Dirac sea (schematic)
Credits: Kim Boström
License: Unrestricted redistribution, commercial use, and modification

Figure 6.5
contains two pictures:
– Foreground: Michelson and Morley at their interferometer
Credits: unknown/Harald Maurer
License: Jede Art von Wiedergabe nur unter Quellenangabe gestattet (Any type of
reproduction is only permitted if the source is mentioned)
Source: http://www.mahag.com/srt/michel2.php
– Background: Molecular vortices
Credits: James Clerk Maxwell / Zwikki at de.wikipedia
License: Unrestricted redistribution, commercial use, and modification
Source: https://commons.wikimedia.org/wiki/File:Maxwell_Molekularwirbel.jpg

Figure 6.6
Black hole Messier 87 Credits: Event Horizon Telescope
License: Creative Commons Attribution 4.0 International
Source: https://commons.wikimedia.org/wiki/File:Black_hole_-_Messier_87_crop_
   max_res.jpg

Figure 6.7
contains two panels:
– Quantum hall effect (measured data) Credits: Antikon at ru.wikipedia
License: Creative Commons Attribution-Share Alike 3.0 Unported
Source: https://commons.wikimedia.org/wiki/File:Quantum_Hall_effect.png

– Klaus-Olaf von Klitzing explaining the Quantum Hall Effect at Georgia Tech
Credits: Ramuman at en.wikipedia
License: Public Domain
Source: https://commons.wikimedia.org/wiki/File:Klausvonklitzing.jpg

References

Abbott, B. P., R. Abbott, T. D. Abbott, M. R. Abernathy, F. Acernese, K. Ackley, C. Adams, et al. 2016. »Observation of Gravitational Waves from a Binary Black Hole Merger.« *Physical Review Letters* 116 (6): 061102.

Akiyama, K., Alberdi, A., Alef, W., Asada, K., Azulay, R., Baczko, A.-K., Ball, D., et al. 2019. »First M87 Event Horizon Telescope Results. IV. Imaging the Central Supermassive Black Hole.« *The Astrophysical Journal* 875 (1): L4.

Corry, L. 1997. »Belated Decision in the Hilbert-Einstein Priority Dispute.« *Science* 278 (5341): 1270–3.

Creţu, A.-M. 2020. »Diagnosing Disagreements: The Authentication of the Positron 1931–1934.« *Studies in History and Philosophy of Science Part B: Studies in History and Philosophy of Modern Physics* 70: 28–38.

Dittrich, Walter. 2015. »On the Pauli-Weisskopf Anti-Dirac Paper.« *The European Physical Journal H* 40 (2): 261–78.

Einstein, A. 1905a. »Über einen die Erzeugung und Verwandlung des Lichtes betreffenden heuristischen Gesichtspunkt.« *Annalen Der Physik* 322 (6): 132–48.

Einstein, A. 1905b. »Zur Elektrodynamik bewegter Körper.« *Annalen Der Physik* 322 (10): 891–921.

Einstein, A. 2007. »Ether and the Theory of Relativity.« In *The Genesis of General Relativity. Boston Studies in the Philosophy of Science, Vol 250*, edited by M. Janssen, J.D. Norton, J. Renn, T. Sauer, & J. Stachel, 1537–42. Dordrecht: Springer.

Ellis, G.F.R. 2014. »Stephen Hawking's 1966 Adams Prize Essay.« *The European Physical Journal H* 39 (4): 403–11.

Ford, L.H. 2003. »The Classical Singularity Theorems and Their Quantum Loopholes.« *International Journal of Theoretical Physics* 42 (6): 1219–27.

Larsson, M. & Balatsky, A. 2019. »Paul Dirac and the Nobel Prize in Physics.« *Physics Today* 72 (11): 46–52.

Lorenz, E.N. 1963. »Deterministic Nonperiodic Flow.« *Journal of the Atmospheric Sciences* 20 (2): 130–41.

Mandelbrot, B.B. 1982. *The Fractal Geometry of Nature*. San Francisco: Freeman.

Maxwell, J.C. 1861. »On Physical Lines of Force.« *The London, Edinburgh, and Dublin Philosophical Magazine and Journal of Science* 21 (141): 338–48.

Maxwell, J.C. 1865. »A Dynamical Theory of the Electromagnetic Field.« *Philosophical Transactions of the Royal Society of London* 155: 459–512.

Michel, J.G. 2019. »How Are Species Discovered? Declarative Speech Acts in Biology.« *Grazer Philosophische Studien* 96 (3): 419–41.

Michel, J.G. 2020. »Could Machines Replace Human Scientists? Digitalization and Scientific Discoveries.« In *Artificial Intelligence: Reflections in Philosophy, Theology, and the Social Sciences*, edited by B.P. Göcke & A. Rosenthal-von der Pütten, 361–76. Paderborn: mentis/Brill.

Pauli, W. 1927. »Zur Quantenmechanik des magnetischen Elektrons.« *Zeitschrift für Physik* 43 (9–10): 601–23.

Planck, M. 1901. »Ueber das Gesetz der Energieverteilung im Normalspectrum.« *Annalen der Physik* 309 (3): 553–63.

Planck, M. 1943. *Wege zur physikalischen Erkenntnis: Reden und Vorträge*. Vol. 1. Leipzig: Hirzel.

Poincaré, H. 1890. »Sur le problème des trois corps et les équations de la dynamique.« *Acta Mathematica* 13 (1): A3–A270.

Poincaré, H. 1900. »La théorie de Lorentz et le principe de réaction.« *Archives Néerlandaises des Sciences Exactes et Naturelles* 5: 252–78.

Poincaré, H. 1904. »L'état actuel et l'avenir de la physique mathématique.« *Bulletin des Sciences Mathématiques* 28 (2): 302–24.

Poincaré, H. 1905. »Sur la dynamique de l'électron.« *Comptes Rendus Hebdomadaires des Séances de L'Académie des Sciences* 140: 1504–8.

Standage, T. 2000. *The Neptune File: A Story of Astronomical Rivalry and the Pioneers of Planet Hunting*. New York: Walker.

Thomson, W. 1869. »On Vortex Atoms.« *Proceedings of the Royal Society of Edinburgh* 6: 94–105.

Thorne, K. 1994. *Black Holes & Time Warps: Einstein's Outrageous Legacy*. New York, London: WW Norton & Company.

Todorov, I.T. 2005. »Einstein and Hilbert: The Creation of General Relativity.« *arXiv:Physics/0504179* [*Physics.hist-Ph*], no. November 1915: 15. https://arxiv.org/abs/physics/0504179.

van der Laan, S. 2013. »The Vortex Theory of Atoms: Pinnacle of Classical Physics.« PhD thesis, Utrecht University.

Veisdal, J. 2019. »Einstein and Hilbert's Race to Generalize Relativity.« https://medium.com/cantors-paradise/einstein-and-hilberts-race-to-generalize-relativity-6885f44e3cbe.

von Helmholtz, H. 1858. »Über Integrale der hydrodynamischen Gleichungen welche den Wirbelbewegungen entsprechen.« *Journal für die reine und angewandte Mathematik* 33: 485–512.

Wikipedia contributors. 2020. »Luminiferous aether.« In *Wikipedia, the Free Encyclopedia.* https://en.wikipedia.org/w/index.php?title=Luminiferous_aether&oldid=955598434.

Scientific Discoveries in Biology

Michael Schmitt

7.1 Introduction

In biology, different kinds of discovery can be made: causations of empirically observed phenomena, law-like relationships in nature, and also species previously unknown to science (see Michel 2019). Such discoveries of »new species« can be made in different ways. The Grand Old Man of German coleopterology, Gustav Adolf Lohse (1910-12-27–1994-04-30), reported regularly to the annual meeting of Central European coleopterists that he had found during the previous year »a dozen of beetle species new for Germany« (or »Central Europe«). He also emphasised regularly that – in spite of the speculations in the audience if he had succeeded to complete the dozen – he always had to decide which species to select for his presentations. »New« species for a given area, e.g. Germany, can come into scientific existence by (1) migration from neighbouring areas, (2) importation through inadvertent human action, (3) purposeful importation, (4) taxonomic action (splitting existing nominal species), (5) detection of a previously overlooked species in the field, or (6 – the least probable) natural speciation. In any of these cases, they can (and must) be discovered in order to be recognised. This opens the question how scientists discover species, or anything else in biology. To get an idea of the process of discovery in biology, I describe five examples: (1) the so-called Darwin-finches and their impact on the development of Charles Darwin's theory of evolution, (2) the viviparous shark of Aristotle and its (re-)discovery by Johannes Müller, (3) the discovery of ultrasound detection in moths by Friedrich Schaller, (4) the discovery of the »new insect order« Mantophasmatodea by Klaus-Dieter Klass and co-authors, and (5) the predicted discovery of the Madagascan hawkmoth *Xanthopan morganii praedicta*.

7.2 Examples

7.2.1 *The Darwin-finches*

Charles Darwin (1809-02-12–1882-04-19) collected 31 specimens of finches while on the Galápagos Islands between 15th September and 20th October, 1835. These finches became icons of evolutionary biology and are presented in literally all popular books and textbooks on evolution as convincing evidence of adaptive radiation (see, e.g., Sulloway 1982a, Milner 1990, 121, Voss 2007, 27ff.). The finder of these birds (Darwin), however, had not even recognised that he had collected birds of different species on different islands of the Galápagos archipelago. Only in 1837, the ornithologist John Gould (1804-09-14–1881-02-03) published original descriptions of these species. Darwin mentioned the finches in the first edition of his scientific report on the voyage of the Beagle only briefly (Darwin 1839, 461f.), and only in 1845 he included the later famous illustration of four species (Darwin 1845, 379, see fig. 7.1). The finches are not mentioned explicitly in any of the editions of the *Origin of Species* (Darwin 1859ff.). According to Sulloway (1982b), it took more than a century until scientists treated Darwin's finches as examples of adaptive radiation (Lack 1947).

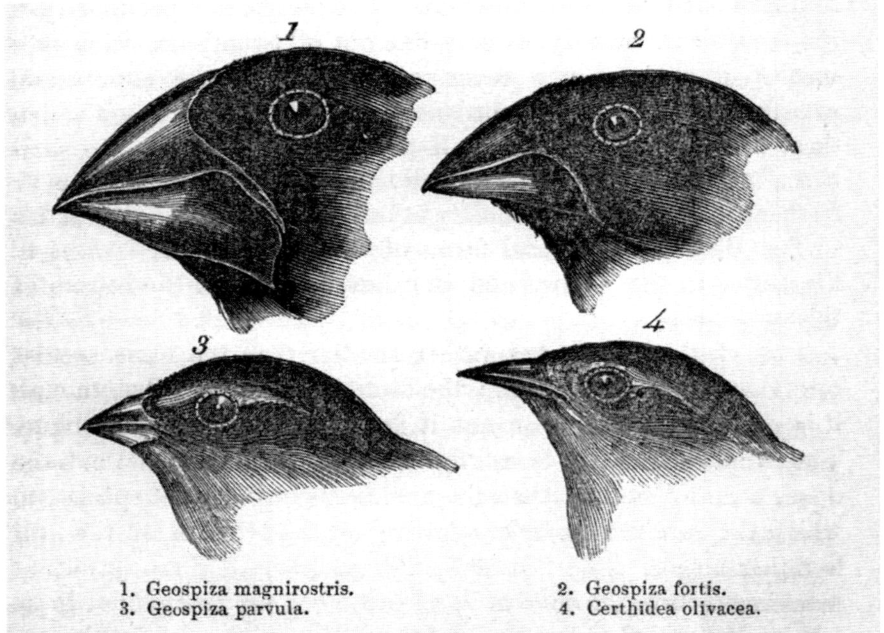

Figure 7.1   The only illustration in a work of Charles Darwin depicting finches collected on the Galápagos Islands (1845, 379).

7.2.2   *Aristotle's Shark*

In his *Historia Animalium*, Aristotle (384–322 B.C.) mentioned (1910, 178) a shark the females of which nurse their young internally by means of a placenta: »The so-called smooth shark has its eggs in betwixt the wombs like the dogfish; these eggs shift into each of the two horns of the womb and descend, and the young develop with the navel-string attached to the womb, so that, as the egg-substance gets used up, the embryo is sustained to all appearance just as in the case of quadrupeds«. Scientists accepted this statement as true during medieval times, without checking its content empirically. Niels Stensen (1638-01-11–1686-12-15) dissected a shark in 1673 and confirmed Aristotle's observation. However, modern scientists forgot about Stensen's report, and Aristotle's description reached the status of a legend (Koller 1958, 117). Johannes Peter Müller (1801-07-14–1858-04-28) took Aristotle's writings for granted and searched intensively for the shark with placenta-forming females. He (re-)discovered this species and identified it as *Mustelus laevis*, now a junior synonym of *Mustelus mustelus* (Linnaeus, 1758). In 1840, Müller reported his discovery to the Prussian Academy of Sciences (published 1842, see fig. 7.2).

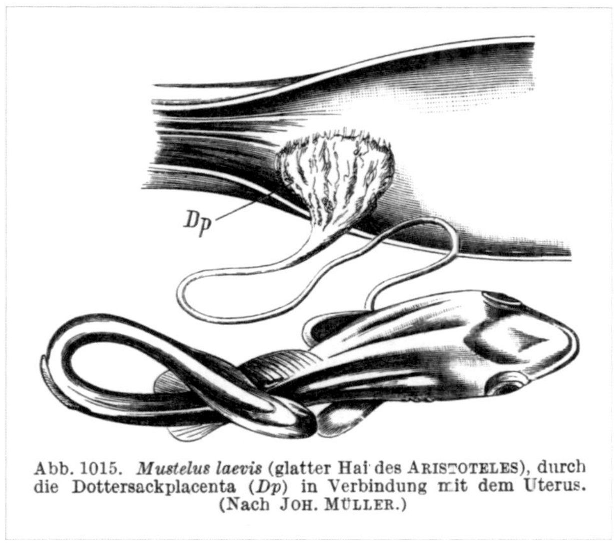

Abb. 1015. *Mustelus laevis* (glatter Hai des ARISTOTELES), durch die Dottersackplacenta (*Dp*) in Verbindung mit dem Uterus. (Nach JOH. MÜLLER.)

Figure 7.2   Illustration of Aristotle's shark in the zoology textbook of Claus et al. (1932, 925), compiled after several original figures from Müller (1842).

7.2.3   *Ultrasound Perception in Moths*

The following section is based on my personal memories and notes of the recollections told by the German-born Austrian zoologist Friedrich Schaller (1920-08-20–2018-05-05), who had given a talk at the university of Freiburg im Breisgau (Germany) on 27th of January, 1984. He was a young university assistant at Mainz university in the late 1940s and interested – among other topics – in sound production and perception in insects (Schaller 2000: 116-119). On a warm summer's evening a group of zoologists, among them (if I recall correctly) also the Dutch zoologist Sven Dijkgraaf (1908-04-17–1995-12-05), had an open air wine party. Out of a sudden, some moths fell from the sky, some of them on the table, crawled around for a while and flew away after a minute or so. This happened several times at irregular intervals of several minutes. Friedrich Schaller was the only one who realised that one of the company at table played around with a cork and a wine bottle. This colleague (possibly S. Dijkgraaf) rubbed the cork up and down over the moist surface of a standing bottle and by doing so produced a faint squeaking sound. Friedrich Schaller observed that the moths dropped always immediately after such a squeaking sound and concluded by inferring *post hoc ergo propter hoc* a causal relation between the sound and the reaction, and he further speculated that it might be the ultrasonic component inaudible to human ears that elicits the reaction of the moths. He subsequently tested this idea in the lab and published the discovery of ultrasound perception in moths (Schaller & Timm 1950). Later research showed that there are several different escape reactions performed by moths perceiving bat echolocation signals (Roeder 1975, see also fig. 7.3).

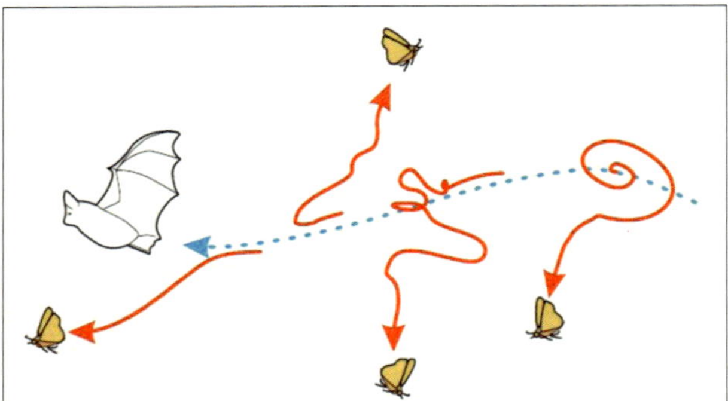

Figure 7.3   Four different escape reactions of moths when perceiving ultrasound echolocating signals of a foraging bat (from Jacobs & Bastian 2016, with kind permission).

7.2.4   *Mantophasmatodea*

When an animal species is discovered that is new to science and does not fit into any of the already described orders, the public attention is immense. So it was when Klass et al. published the name of a new order for a recently discovered insect species (Klass et al. 2002). It might be useful to the non-biologist readership to know that in biological, especially zoological, Linnaean encaptic classification species are grouped into genera, genera into families, families into orders, orders into classes etc. This means that an »order« is a category of relatively high rank, and consequently a »new order« attains much more public interest – within and outside science – than just another new species. Interestingly, individuals of this and other similar species were known for a long time, for example to the South African entomologist Mike Picker (see Eberhard 2009, fig. 7.4). However, entomologists who knew these critters deemed them being wingless juveniles of some strange grasshoppers and did not realise that they represent a »new order«.

Figure 7.4   *Karoophasma biedouwense* male
              (Mantophasmatodea); photo by Monika Eberhard,
              with kind permission.

During his doctoral project, Oliver Zompro discovered in a museum collection some amber fossils of insects that he could not sort into accepted species, genera, families or even orders. In addition, he studied material of extant species and realised that also these specimens did not fit in already described taxonomic units. Therefore, he sought the advice of more experienced entomologists, his doctoral supervisor, Joachim Adis (1950-03-04–2007-08-29), the morphologist and phylogeneticist Klaus-Dieter Klass, and the Danish entomologist Niels Peder Kristensen (1942-03-02–2014-12-06) who certainly had the broadest experience in insect taxonomy at that time. Only in a joint effort and after pondering all alternative solutions, they published their discovery of a new insect order, the Mantophasmatodea. Although the entomological community debated for some time the erection of a new order for the newly discovered – or rather recognised – species, the fact of the discovery as well as the new scientific name, and also the vernacular names »gladiators« or »heelwalkers«, were soon accepted publicly. Meanwhile, two gladiators are depicted

on a postal stamp of the Republic of South Africa, and models made of silver
thread and glass beads were sold as souvenirs of the 23rd International Congress
of Entomology, held in Durban in 2008 (fig. 7.5). The subject of the discovery was long known to science but misinterpreted. Two of the specimens that
Oliver Zompro studied were collected in 1909 and 1950 and remained unrec-
ognised since then on the shelves of the museums in Berlin and Copenhagen.

Figure 7.5   Mantophasmatodea as souvenir and on a stamp
               (originals).

7.2.5   *The Madagascan Hawkmoth*

Charles Darwin had learned of an orchid, *Angraecum sesquipedale*, on the
island of Madagascar, the flowers of which had nearly 11.5 inches (30 cm)
long nectaries or labiate spurs, where only the last 1.5 inches are filled with
nectar. He stated in his treaty »On the various contrivances by which British
and foreign orchids are fertilised by insects« (1862, 198) that there must be a
moth with a proboscis that is between 10 and 11 inches long so that the moth
can access the nectar when visiting and pollinating the orchid flower. Such
a moth was not known at that time but was discovered only more than forty
years later (Rothschild & Jordan 1903). The first describers named this moth
Xanthopan morganii praedicta, with reference to Darwin's prediction. The species *Xanthopan morganii* belongs to the family of hawkmoths (Sphingidae)
and was known to occur also on the mainland of Africa, but only the individuals of the Madagascan population (subspecies *Xanthopan morganii paedicta*)
have such long probosces (fig. 7.6).

Figure 7.6   A *Xanthopan morganii praedicta* hawkmoth visiting an *Angraecum sesquipedale* flower (https://www.britannica.com/plant/Angraecum-sesquipedale#/media/1/25242/157393).

7.3   Discussion

In an attempt to characterise the essentials of a scientific discovery, at least in biology, it is useful to distinguish between a »finding« and a »discovery« proper. The *Oxford English Dictionary* (cited after the online edition) defines

> **Finding:** The action of coming across or discovering something or someone by chance or as the result of searching or enquiry; an instance of this.

> **Discovery:** (a) The action of finding out or becoming aware of something for the *first time*; the action of being the first to find (a place); the action of bringing to light something (as a substance, scientific phenomenon, etc.) which was previously unknown. [...] Something which is discovered, revealed, or brought to light; a substance, phenomenon, etc., found *for the first time* [emphases added].

In empirical sciences (e.g. in biology), finding something is a normal, even an essential component of daily work. Sometimes, such a »normal« finding is regarded a »discovery«. Sometimes, it takes some time until a ›finding‹ turns into a ›discovery‹, as with the Darwin finches. Some discoveries are made after intentional search, as with Aristotle's shark. Others are made by chance, as the ultrasound perception in moths. But this case demonstrates that a certain background knowledge is a prerequisite of discovering something: Of all the members of the company at table only Friedrich Schaller combined his observations – moths falling from the sky after someone had produced a certain noise – with his idea that moths are able to perceive the echolocation sounds of foraging bats. How fast the scientific community accepts a discovery depends, among other factors, also on the circumstances of its publication. Gregor Mendel (1822-07-22–1884-01-06) published his discovery of the laws of heredity in a journal of very limited outreach (Mendel 1866). He was recognised as the founder of a scientific revolution not until 16 years after his death (Janko & Matálova 2001, Magner 2002, 376-391). A more recent case of delayed public acceptance is the discovery of the transposons (*vulgo* jumping genes) by Barbara McClintock (1902-06-16–1992-09-02) in 1951, for which she received the Nobel prize for Medicine and Physiology only in 1983. Even if the importance of her discovery was realised several years earlier, the delay between the original publication in 1951 and its public acceptance is remarkable (Keller 1995, Schmitz 2001).

The discovery of a new insect order by Klass et al. in 2002 is an example of an entirely different style of publication. The authors placed their paper in a scientific journal with high impact. They made clear from the beginning that they did not just discover some species that were previously not known to science but emphasised that it was necessary, or at least justified, to erect a new »order« to place these species in the tree of life. Their paper was cited 203 times until now according to Google Scholar (5th May, 2021), which is comparably high for a taxonomic paper. Klass et al.'s discovery shows that discoveries in biology, especially in taxonomy, can be made on material collected long before and stored for decades on the shelves of museums (Zompro et al. 2002). Fontaine et al. (2012) found that the average shelf-life of 570 randomly taken species from the ca. 16.000 described in 2007 was 20.7 years. Such findings resemble the discovery of an overlooked item in an archive or a library.

There are certainly numerous cases of foreseeable discoveries in biology. Quite a striking one is the example of »Darwin's Madagascan hawk moth prediction« (Kritsky 1991). It demonstrates that Darwin's understanding of the pollination of orchids and his knowledge of possible pollinators was decisive for his insight that there must be a moth with a more than 25 cm long proboscis in Madagascar. This and the other aforementioned cases show that discoveries in biology can only be made if the discoverers have a certain background knowledge, and that credit is given to the original authors even if the public acceptance of such a discovery might be delayed.

In fig. 7.7 I list a putative array of terms and notions coming to my mind when I aimed at describing »scientific discovery«. It seems to me that finding something unexpected, the »eureka-moment«, is crucial for counting a finding as a discovery.

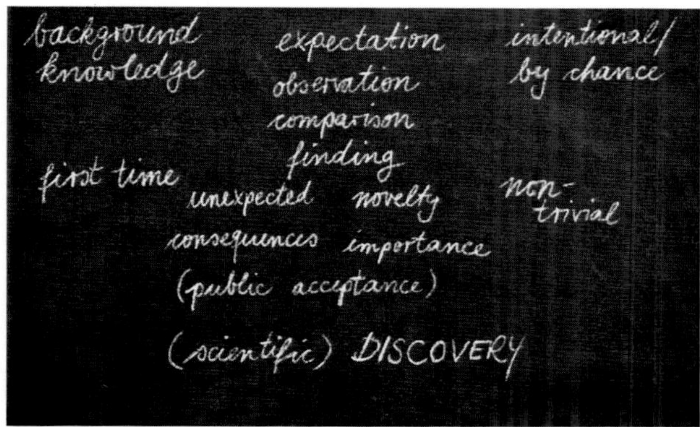

Figure 7.7   Notions from a mind map pertaining to »scientific discovery« in a putative sensible order (original, thanks to Jan Michel).

As I see it, we – scientists – develop an expectation of what we shall find, based on our background knowledge in the field of our investigations. Either intentionally or by chance we make an observation that we compare to our expectation and find that the observation either matches our expectation or not. As Popper (1994, 16) put it, a failed expectation is the starting point of a search for a solution. In case of non-match, it could be that we found something for the first time. Finding something unexpected is an important concomitant of a »discovery«, but the moment of surprise occurs possibly only to a slight degree – if the discoverer had a clear expectation of making the discovery, as, e.g., Johannes Müller when he searched for Aristotle's shark. This shows that the discoverer him- or herself may be hardly surprised because he or she expected the finding. The surprise can be much more on the side of the public than of the discoverer. Nevertheless, if something is observed totally in agreement with our expectation, we would hardly count it as a discovery. However, if the finding is a true novelty and non-trivial, it will have important consequences for the subsequent development of the respective field of science. And only in this case the scientific community will accept this finding as a discovery.

The weight of »public acceptance« for the status of a ›finding‹ as a ›discovery‹ is variable and relative. What could cause an upset in a small circle of scientific specialists could be regarded irrelevant for a larger community of generalists or a similarly small circle of specialists in a different field. Also, what is ranked as an important discovery might be treated as trivial a century later or among members of a different community. In one circle, detecting a new species is regarded a »discovery« at any rate, even if it is the 457th species of rover beetles of a certain region. Others might find that calling this a »discovery« is an inflationary use of this term. Even among the coleopterists mentioned in the introduction, not all twelve discoveries of »new species« were viewed as equally important. It makes certainly a difference whether a species is counted as new for a certain area because it was introduced inadvertently or if it was not known to science at all. This simply means that science is not done in a sterile environment, but that the social, economic, and political context matters. Even gender affairs play a considerable part, as in Barbara McClintock's case.

A probably underestimated component of ›scientific discovery‹ is good luck. It is impossible to estimate how many scientific discoveries were not made due to the lack of at least a pinch of good luck. Finally, as Sidney John and Caroline Hutt put it (Hutt & Hutt 1970, VIII), »we see only what we look for, and we look for only what we know, for we are creatures that look before and after and occasionally around to really see what is passing under our very eyes. The empiricist thinks he believes only what he sees, but he is much better at believing than at seeing [... .]«.

Acknowledgements

I thank Jan Michel (Bochum/Düsseldorf, Germany) for the invitation to participate in the symposium Making Scientific Discoveries held at Münster on 3–5 December, 2019, him, Michael Ohl (Berlin, Germany) and Gabriele Uhl (Greifswald, Germany) for stimulating discussions and valuable feedback, and Monica M. Sheffer (Greifswald, Germany) for carefully checking and improving the English of my text.

References

Aristotle [1910]. *History of Animals* (Περὶ τὰ ζῷα ἱστορίαι). Cited after https://www.globalgreyebooks.com/history-of-animals-ebook.html.

Claus, C., Grobben, K., & Kühn, A. 1932. *Lehrbuch der Zoologie, Spezieller Teil.* Berlin: Springer. Reprint 1971.

Darwin, C. 1839. *Journal of Researches into the Geology and Natural History of the Various Countries Visited by H.M.S. Beagle.* London: Colburn. http://darwin-online. org.uk/converted/pdf/1839_Beagle_F11.pdf.

Darwin, C. 1845. *Journal of Researches into the Geology and Natural History of the Various Countries Visited by H.M.S. Beagle Round the World.* 2nd ed. London: John Murray. http://darwin-online.org.uk/content/frameset?itemID=F14&viewtype=image&pageseq=392.

Darwin, C. 1859. *On the Origin of Species by Means of Natural Selection, or the Preservation of Favoured Races in the Struggle for Life.* London: John Murray. http://darwin-online. org.uk/EditorialIntroductions/Freeman_OntheOriginofSpecies.html.

Darwin, C. 1862. *On the Various Contrivances by Which British and Foreign Orchids Are Fertilised by Insects.* London: John Murray. http://darwin-online.org.uk/ EditorialIntroductions/Freeman_FertilisationofOrchids.html.

Eberhard, M. J. B. 2009. »Kurze Vorstellung der Ordnung Mantophasmatodea.« *Entomologica Austriaca* 16: 73–84.

Fontaine, B., Perrard, A., & Bouchet, B. 2012. »21 Years of Shelf Life Between Discovery and Description of New Species.« *Current Biology* 22 (22): R943–R944. doi. org/10.1016/j.cub.2012.10.029.

Gould, J. 1837. »Remarks on a Group of Ground Finches From Mr. Darwin's Collections, With Characters of the New Species.« *Proceedings of the Zoological Society of London* 5: 4–7. https://www.biodiversitylibrary.org/item/96163#page/176/mode/1up.

Hutt, S.J. & Hutt, C. 1970. *Direct Observation and Measurement of Behavior.* Springfield, Ill.: Charles Thomas.

Jacobs, D.S. & Bastian, A. 2016. »Passive and Active Acoustic Defences of Prey Against Bat Predation.« In *Predator–Prey Interactions: Co-Evolution Between Bats and Their Prey*, 43–71. Cham: Springer. doi: 10.1007/978-3-319-32492-0_4.

Janko, J. & Matálová, A. 2001. »Johann Gregor Mendel (1822–1884).« In *Darwin & Co.: Eine Geschichte der Biologie in Portraits*, vol. 1, edited by I. Jahn & M. Schmitt, 390–410, 533–534. München: C.H. Beck.

Keller, E.F. 1983. *A Feeling for the Organism: The Life and Work of Barbara McClintock.* New York: Freeman.

Klass, K.-D., Zompro, O., Kristensen, N.P., & Adis, J. 2002. »Mantophasmatodea: A New Insect Order with Extant Members in the Afrotropics.« *Science* 296: 1456–1459. https://science.sciencemag.org/content/296/5572/1456.

Kritsky, G. 1991. »Darwin's Madagascan Hawk Moth Prediction.« *American Entomologist* 37 (4): 206–210. https://doi.org/10.1093/ae/37.4.206.

Lack, D. 1947. *Darwin's Finches: An Essay on the General Theory of Evolution.* Cambridge: Cambridge University Press.

Magner, L.N. 2002. *A History of the Life Sciences*. 3rd ed. New York: Marcel Dekker.

McClintock, B. 1951. »Chromosome Organization and Genic Expression.« *Cold Spring Harbor Symposia on Quantitative Biology* 16: 13–47. doi: 10.1101/SQB.1951.016.01.004.

Mendel, G. 1866. »Versuche über Pflanzenhybriden.« *Verhandlungen des naturforschenden Vereines in Brünn, Bd. IV für das Jahr 1865:* 3–47. https://www.zobodat.at/pdf/Flora_89_0364-0403.pdf.

Michel, J.G. 2019. »How Are Species Discovered? Declarative Speech Acts in Biology.« *Grazer Philosophische Studien* 96 (3): 419–441. doi 10.1163/18756735-09603011.

Milner, R. 1990. *The Encyclopedia of Evolution: Humanity's Search for Its Origins*. New York: Facts on File.

Müller, J. 1842. »Über den glatten Hai des Aristoteles, und über die Verschiedenheiten unter den Haifischen und Rochen in der Entwicklung des Eies.« *Physikalische Abhandlungen der Königlichen Akademie der Wissenschaften zu Berlin* 1840: 187–257. https://www.biodiversitylibrary.org/item/92569#page/231/mode/1up.

Oxford English Dictionary, online version (https://www.oed.com/).

Popper, K.R. 1994. *Alles Leben ist Problemlösen: Über Erkenntnis, Geschichte und Politik*. München, Zürich: Piper.

Roeder, K.D. 1975. »Neural Factors and Evitability in Insect Behavior.« *Journal of Experimental Biology* 194 (1): 75–88. doi.org/10.1002/jez.1401940106. PMID 1194872.

Rothschild, W. & Jordan, K. 1903. *A Revision of the Lepidopterous Family Sphingidae*. *Novitates Zoologicae* 9, supplement. London & Aylesbury: Hazell, Watson & Viney. https://doi.org/10.5962/bhl.title.5651.

Schaller, F. 2000. *Erfüllte Endlichkeit: Autobiografie des Zoologen Friedrich Schaller*. Linz: Oberösterreichisches Landesmuseum, Biologiezentrum.

Schaller, F. & Timm, C. 1950. »Das Hörvermögen der Nachtschmetterlinge.« *Zeitschrift für vergleichende Physiologie* 32: 468–481. https://link.springer.com/article/10.1007/BF00339923.

Schmitz, S. 2001. »Barbara McClintock (1902–1992).« In *Darwin & Co.: Eine Geschichte der Biologie in Portraits*, vol. 2, edited by I. Jahn & M. Schmitt, 490–505, 562. München: C.H. Beck.

Sulloway, F.J. 1982a. »Darwin and His Finches: The Evolution of a Legend.« *Journal of the History of Biology* 15 (1): 1–53.

Sulloway, F.J. 1982b. »Darwin's Conversion: The *Beagle* Voyage and Its Aftermath.« *Journal of the History of Biology* 15 (3): 325–396.

Voss, J. 2007. *Darwins Bilder: Ansichten der Evolutionstheorie 1837–1874*. Frankfurt: Fischer Taschenbuch.

Zompro, O., Adis, J., & Weitschat, W. 2002. »A Review of the Order Mantophasmatodea (Insecta).« *Zoologischer Anzeiger – A Journal of Comparative Zoology* 241 (3): 269–279.

Discoveries in Linguistics: Making, Finding, or Somewhere in Between?

Mitchell S. Green

8.1 Some Intuitions about Scientific Discovery

Discovery is crucial to the progress of science.[1] At a minimum, a scientific discovery involves coming to know something that was not known previously. Because discovery is an epistemic phenomenon, it is also factive. One cannot therefore discover that human beings evolved from other non-extinct great apes, or that we only use 10% of our brains. Further, even if she ends up being proven correct, an investigator who makes an educated guess or conjecture does not, at least on this basis, merit being called a discoverer of a phenomenon.[2]

Scientific discovery's epistemic character also enables us to see that it requires conceptual sophistication needed to grasp the phenomenon being discovered. This is why, although an undergraduate new to chemistry might synthesize a novel compound by a fortuitous and largely random addition of ingredients, we would not say that he has (scientifically) discovered that new compound. Likewise even if a (non-savant) toddler happens upon a species of butterfly unknown to biology, she has not made a scientific discovery.

The picture of science bequeathed to us by the Enlightenment is that of a lone investigator delving into a topic that has captured her or his curiosity. On occasion the investigator might report their results to a learned body, but for the most part such work is solitary. By contrast, nowadays great swathes of science are only possible with the collaboration of large groups of researchers, often distributed across the globe, and each performing distinctive tasks relative to the rest of the team. Manuscript and grant referees, journal editors, and institutional review boards are integral to the process as well. This picture might suggest that the process of scientific discovery is now, even if it has not always been, an inherently social one.

---

1   An earlier version of this paper was presented at the Making Scientific Discoveries Conference, Münster, Germany, December, 2019. I am grateful to the audience on that occasion for their perceptive comments. Thanks also to Jan Michel for comments on an earlier draft.

2   This distinction helps to explain why, although we credit Kant with articulating a nebular hypothesis for the formation of stars and planets, historians of science are not prepared to say that he discovered this phenomenon. See for instance Schönfeld 2006.

© BRILL MENTIS, 2022 | DOI:10.30965/9783957437044_009

Scientific discovery is a social phenomenon in at least the following respect. Everyday thought and talk distinguishes between what an individual knows and what is known, where the latter notion concerns what is known by the community. Following Hyman (1999), we may accordingly distinguish between personal and impersonal knowledge.[3] This distinction correlates with two notions of discovery. An investigator might feel that she has discovered fact F without thinking she has made an original contribution to scientific knowledge. Even if she has not made an original contribution, she may have made a personal discovery. But when we use the phrase, ›scientific discovery‹, we tend to mean that someone or a group have happened upon new information that is of interest to the scientific community, at least in part because it is not yet impersonally known.[4]

We may be tempted to take a further step and suggest that to be a scientific discovery, an epistemic advance must become part of what is impersonally known by being accepted as fact by a sufficient number of, or appropriately placed, members of the scientific community, or at least of the sub-community concerned with the topic in question. Short of that, the epistemic advance will at best produce one or more instances of personal knowledge. In spite of its attraction, however, we should be wary of this step. The reason is that it fails to respect the mundane observation that someone could discover a fact that is not impersonally known and that, if it were impersonally known, would advance scientific understanding; and yet her discovery goes unacknowledged due to an untimely death or the machinations of a repressive, anti-scientific regime. A lone spelunker happens upon a cave that contains exquisitely preserved drawings by pre-historic humans; but soon thereafter a massive earthquake kills the spelunker and destroys the cave complex along with all the artwork it contains. Or an amateur astronomer discovers a comet previously unknown to mankind; but since recognition of that comet would undermine the authority of the local despot, the astronomer and all her instruments and

---

3   Hyman in turn draws on Williams 1972. The notion of impersonal knowledge raises intriguing questions, such as: when an isolated community knows something that the rest of humanity does not, is it impersonally known? Also, to be impersonally known, must the group members in question be aware of one another's epistemic states? Fortunately, we do not need to settle these questions in order to make progress on the issues that form the main concern of this essay, and they will not be pursued here.

4   It will not be necessary for the discussion that follows to clarify the distinction between contributions to scientific knowledge and to other kinds of impersonal knowledge, such as history. However, we might expect that contributions to scientific knowledge must add to the development of one or more theories that are suited for both prediction and explanation, while contributions to other kinds of impersonal knowledge are not put up to this standard.

data are liquidated. In both cases we would seem to have discoveries – indeed discoveries of interest to science – that are never acknowledged by others.

A similar picture would seem to apply not just to archeology and astronomy, but also to the field of linguistics, which will be our focus in what follows. More generally, our naïve thinking about discoveries in linguistics would lead us to suppose that when a linguist (professional or amateur) makes a discovery in her field, the following conditions are met:

1.   There is a fact F pertaining to language, such as what words are or are not possible given the facts of acoustics and human physiology, or a general phenomenon of language change, such as that governing vowel shift or the progression from pidgins to creoles.

2.   Fact F is not currently impersonally known.

3.   The investigator employs both conceptual and communicative skills needed not only to isolate fact F, but also to describe it accurately.

4.   The investigator employs these skills in such a way as to state and justify her claim that fact F obtains.

5.   Typically, the investigator communicates her new discovery to the appropriate linguistic experts, such as morphologists or historical linguists, who after refereeing and other mandatory vetting, accept this new finding in such a way as that it now becomes impersonal linguistic knowledge.

6.   Typically, a newly discovered phenomenon or entity is given a name, sometimes in honor of those who were instrumental in its identification (as we see with the names for Broca's and Wernicke's areas), but more often not (as we see with terms such as ›agglutinative language‹ and ›critical period‹). In both types of phenomena, terms are stipulated to refer to biological structures or linguistic phenomena, though of course these phenomena would exist, and could even have been discovered, without being given an official label.

It follows from the above that two individuals or groups could simultaneously discover a fact F. This in turn implies that an individual or group could discover fact F without being the discoverer of fact F. Further, and keeping in mind that important discoveries are now regularly made by large teams of investigators, we would do well also to countenance a seventh feature of the discovery process, namely:

7.   The linguistic discovery might be achieved by a large group G of investigators, where each member of the group may have distinctive skills and knowledge. In particular, the intellectual achievement might be so complex that no one individual can be said to understand the newly discovered phenomenon completely. Here, we might say that group G has discovered fact F, even though no member of G has done so.

8.2   Scientific Discovery and Institutional Reality

Conditions (1)–(7) seem to capture an intuitive view about discovery in linguistics, albeit one refined with acknowledgment of the division of intellectual labor in some discoveries. Our question is whether condition (5) should be strengthened so as to hold that for an epistemic achievement to be a case of scientific discovery, it must be publicized and subsequently accepted by the scientific community in such a way as to become part of impersonal knowledge.

In recent work Michel (2019) has argued for precisely this conclusion. His discussion focuses on the discovery of new species in biology. However, his arguments do not appear to depend on any premises that apply uniquely to biology, and thus if they are sound, they should generalize.[5] As indicated above, for this paper I will focus on the case of discoveries in linguistics.

To motivate his way of thinking of scientific discovery, Michel expounds some ideas from John Searle's work in the philosophy of language and philosophy of social science. First, Searle invokes the notion of a *declarative speech act*, which is a type of speech act in which an agent or agents make something to be the case by representing it as being the case (Searle 1989). For instance, a duly empowered person or group may declare a certain proceedings open, by such an act as saying, »The proceedings are now open.«[6] By representing the proceedings as being open, the appropriately empowered speaker makes it the case that they are open.[7]

On this basis we have a notion of an *institutional fact*, which is a type of fact that can only exist by virtue of agents declaring it to be so. A metal's having a certain atomic structure is not an institutional fact, as it has that structure regardless of what agents take it to have. On the other hand, being vice-president of a corporation is an institutional fact, as is the property of being a €10 note (as opposed to being a piece of paper with certain markings). Also, although institutional facts can only exist by virtue of declarations, those declarations need not be made explicitly. Even if no one ever explicitly declares all

---

5   Michel also encourages this expectation, writing, »I will only focus on discoveries in biology here, even though I am convinced that the overall argument of this paper can be extended and is applicable to discoveries in other scientific disciplines as well« (2019, 420–1).

6   I take Searle's notion of a declarative speech act to exclude proleptic utterances, in which someone might try to bring about an effect by representing that effect as going to be brought about. Thus a parent might say to her teenager about to go out with friends, »I know that you won't be drinking and driving,« intending that the teen will avoid drinking while driving at least in part as a result her parent's utterance. Unlike proleptic utterances, Searle takes his declarative speech acts to bring about their results constitutively rather than causally.

7   Searle 2010, 35.

spherical pieces of amber to be worth a bushel of wheat, they might still come to have that value if they are treated as having that value by the community in which they are used. Here it would be an institutional fact that all spherical pieces of amber are worth one bushel of wheat even though no explicit declaration ever made it so.

Michel aims to establish the interesting and controversial thesis that being a scientific discovery is a matter that is at least in part an institutional fact. To motivate this viewpoint he asks us to consider three examples. In the first, Albert discovers a hole in his sock (Michel 2019, 416–7). Michel describes this as an event that Albert performs in solitude; our calling it a discovery presupposes that Albert did not know about the hole before this moment; and the discovery itself requires a minimum of conceptual machinery on Albert's part: he needs at the very least to have concepts of socks, holes, and personal property. Further, few would be inclined to suppose that Albert's discovery is an institutional fact: it occurs whether or not anyone (including Albert) declares Albert's finding to be a discovery. At the same time, we would not be inclined to call this case a scientific discovery. The reason is evidently that the fact discovered by Albert is too mundane to count as a contribution to science since it is of no theoretical significance.

In the second case that Michel considers, Blackbeard discovers, with the aid of a map that has come into his possession and a crew of fellow sailors ready to follow him, a treasure on an island to which they journey while following the map's directions (ibid., 427–8). Michel notes that it is difficult to decide whether Blackbeard himself discovered the treasure, or instead whether he and his crew did so. So too, unlike in the case of Albert, Blackbeard had some inkling of the treasure's existence and location before finding it. Again, however, even if we agree that the Blackbeard case is one of discovery, it is not one of scientific discovery. Although the treasure is of great importance to Blackbeard and friends, its discovery is still mundane from the point of view of science.

The third case concerns Charles, a respected academic ichthyologist. One day on vacation near the seashore he encounters a group of fish that he suspects have not yet been identified by science. After careful investigation of sample organisms and relevant published research, Charles writes up his finding in article form and submits it, together with a proposed species name, to an academic journal. The paper is then subject to peer review, after which it is accepted for publication and then duly published. Michel writes,

> By submitting the manuscript to a respected biological journal, Charles declares his findings a new biological species discovery. Several weeks later, the

> peer-review process including Charles' revisions of his paper is over and Charles'
> paper is published. This is a case of success: The scientific community accepts
> Charles' proposal, his findings count as a species discovery new to biology and
> Charles counts as the discoverer of this species. (2019, 428)

On Michel's analysis, Charles' submission is a declarative speech act in the
sense earlier elucidated. That declaration will not, however, guarantee anything on its own. Rather, what is required in addition is an appropriate form
of uptake on the part of the scientific community. Such uptake, Michel avers,
comes in the form of acceptance and perhaps also publication in a respectable
scientific journal, with the implicit acknowledgment of the intellectual value
of Charles' contribution that such acceptance and publication normally carries with it.

According to Michel, then, Charles' findings and subsequent publication
thereof count as a discovery in roughly the same way that, say, crossing a certain line on a playing field with a ball in one's hand counts as a touchdown in
American football, or pushing a certain button in a polling booth counts as
voting for a candidate for public office. In both cases, the activities are possible
only in the presence of particular conventions in a certain community: scor-
ing a touchdown is only possible if appropriate rules of that game are in force
in the relevant community; otherwise all one has done is cross a line with a
funnily-shaped ball. So too, an intellectual achievement is not a scientific discovery until it has been ratified as such by the scientific community.

Michel crystallizes his position as follows:

> Charles' discovery of a new biological species has to be accepted as such by a
> relevant scientific community. Otherwise, it would not be a case of a scientific
> discovery but rather, just like Albert's case, a case of mere finding, an atomic (or
> molecular) discovery. (2019, 429)

Michel's position is just as one would expect given the analogy he wishes to
preserve with institutional facts. Appointing someone, placing a bet, and other
speech acts require uptake of the appropriate kinds, while for Michel, uptake
for putative scientific discoveries takes the form of authorized bodies such as
journals and professional organizations accepting them as scientific discoveries, rather than as mere hypotheses or null results.

In spite of its intriguing affinity with theories of institutional reality, we
should note that Michel's position runs counter to our intuitions about scientific discovery collected together in Section I. For we are here being told that a
necessary condition of being a scientific discovery (as opposed to what Michel
calls an »atomic« or »molecular« discovery), is that someone's work has to be

accepted as such by a relevant scientific community, which in the present case is presumably the ichthyological community. What is the difference between an atomic or molecular discovery on the one hand, and a scientific discovery on the other? Michel answers as follows:

> While Albert's discovery is unstructured (atomic) or minimally structured (molecular), Charles' discovery reveals a certain internal structure with certain structural features. Among these features are the following: having a prior body of biological knowledge, the original finding of the exemplars, the preparation of a manuscript, and the acceptance of the finding as a new biological discovery by a scientific community. (Ibid., 429)

Michel's reasoning seems to be as follows. The reason why Charles' finding must be accepted as such by a relevant scientific community is that unless that occurs, it will fail to have certain structural features. Those structural features include the finder's having the appropriate expertise, the presentation of his findings in manuscript form, and the acceptance of his finding by a scientific community.

Obviously this is not too helpful. We want to know what separates scientific discoveries from other kinds. The answer is that only the former have the right kind of structure. And what kind of structure is that? It's the kind of structure one gets from *inter alia* the requirement of acceptance as such by a scientific community.

None of this shows that Michel's position is incorrect, but we do well to listen to any skeptical challenges it may raise. We begin to appreciate the controversial nature of Michel's position by asking the simple question, whether it might be possible for a person to make a scientific discovery that goes unacknowledged by the scientific community. Here just imagine all the elements of Michel's example of Charles the ichthyologist, minus the final two steps of having his paper accepted and published in an icthyological journal. Imagine instead that Charles mistakenly sends the paper to an address for an *ornithology* journal, which duly rejects it for the reason that it is outside the journal's scope. Not noticing his mistake, Charles receives word of his rejection, falls into a major depression, vows never to study fish (or any other living creature) again, destroys his manuscript, and dies in penurious obscurity.

Michel is committed to holding that in the revised case, Charles has not made a scientific discovery, but at best a molecular discovery. This, however, seems controversial. For it would appear at least as plausible to describe the situation as one in which Charles makes a scientific discovery that goes unacknowledged by the scientific community. The suggestion is of course compatible with the fact that Charles' is an unusual and unfortunate case. No doubt, many and perhaps most scientific discoveries are acknowledged and ratified by the relevant expert community. Notice also that being declared a scientific discovery by a relevant expert community is not sufficient for being such a discovery. The reason is that, bearing in mind that ›discovery‹ is a factive term, we may readily imagine a case in which the community is in error in deeming something to be a discovery: they thought it was a discovery but it was not. (A well-known example is the hoax surrounding Piltdown Man (Russell 2003).)

Michel could revise his strong position to one that is weaker, according to which, to be a scientific discovery, a contribution must be capable of being accepted by the (or an) appropriate validating authority. Such a concession would amount to a more defensible necessary condition on scientific discovery than actual acceptance, and yet it is not empty: if an inquirer takes notes in support of a putative discovery, but such notes are undecipherable even after the best efforts of researchers and forensic experts, and neither the researcher nor any other supporting evidence is available to help clarify her line of thought, then we may reasonably doubt whether any discovery has been made. However, an epistemic achievement may be intelligible in the present sense without requiring any declarative speech act to validate it. Instead, it may be vouchsafed by the truth of the following counterfactual: were the relevant experts to assess the putative discovery in an unprejudiced way, they would likely accept that this discovery is a contribution to scientific knowledge. Such a counterfactual conception of scientific discovery seems more defensible than Michel's. However, such a conception does sever any close connection between scientific discovery and institutional reality.

More generally, I suggest that to be a scientific discovery, a contribution must be intelligible and acceptable to an appropriate audience with the expertise needed to carry out its validation. On this more conservative understanding, a scientific discovery does require the realization of a new fact not hitherto acknowledged by the scientific community; this realization in turn requires that the discoverer has appropriate conceptual resources to come to this realization, and applies both those resources and relevant evidence to advance what we know, rather than just what we hypothesize or conjecture. But if this is correct, then the connection with Searlean institutional reality has been severed. Making a scientific discovery is more like driving than it is like checkmating: even though driving is a highly regulated social practice requiring conceptual sophistication on the part of its practitioners, it can occur in the absence of any social institutions that might regulate it.[8]

---

8 In this respect I agree with Searle, who writes, »... the test for whether a noun makes reference to an institution is whether under that description the object named has deontic

What we have seen thus far leaves room for one dimension of institutional reality in scientific discovery, namely the process not of identifying but rather of naming phenomena that are new to science. In the tragic version of Charles' discovery scenario imagined above, the species he identifies is not given a traditional taxonomic name. As with the naming of a newborn child, or exercising one's »naming rights« to a building one has just endowed, the naming of a new biological species is unquestionably a declarative speech act in the sense adumbrated above, and so will create a new institutional fact when carried out successfully. There is, further, no reason that matters should be different in other areas of scientific inquiry. However, discovering a phenomenon, scientific or otherwise, and naming that phenomenon, are quite different matters. For this reason, we may acknowledge the importance of naming phenomena that are new to science, without needing to concede any ground to Michel's argument for the institutional character of scientific discovery.

8.3   Three Case Studies from Linguistics

As users of language, human beings create the facts that linguistics describes. For this reason, one might expect the science of linguistics to provide comparatively favorable cases in which to test the claim that scientific discoveries are institutional facts. I shall do so with reference to three cases studies: Universal Grammar, classifier languages, and Horn scales.

8.3.1   *Universal Grammar*

Universal grammar (hereafter UG) has »emerged as the most prominent hypothesis in the past 50 years on the nature of language.« (Everett 2012, 555) Some of its principal tenets are as follows:

> 1. All normal members of our species are competent in at least one language having the properties of recursive syntax and semantics.
> 2. This competence is arrived at in a way that is driven genetically in the following sense: so long as a child is exposed to language from an early age, she will acquire a competence with that language without needing explicit instruction from someone already possessing such competence.
> 3. The recursive nature of the syntax of a language implies that there is no finite upper bound on the length of sentences that are grammatical in that language.

---

powers. On this test, the Catholic Church is an institution; religion is not. The National Science Foundation is an institution; science is not. Private property is an institution; a car is not.« (2010, 92).

UG is hypothesized to explain such phenomena as the striking rapidity with which children learn the grammar of the languages in which they are raised, and proponents of UG contend that this remarkable learning ability could not be accounted for on the basis of general intelligence alone. For this reason, proponents of UG hold that the process of acquiring a language is different in kind from the process of acquiring knowledge of colors or of different breeds of domestic dog.

In a challenge to UG, Everett (2012) summarizes a series of studies of the South American language Pirahã, which he claims not to be recursive. For instance, Everett reports that his informants who are native speakers of Pirahã will judge the following sentence to be grammatical:

Xahoapioxio	xigihí	toioxaaga	hi	kabatií	xogií	xi
Another day	man	old	he	tapir	big	it
mahahaíhiigí	xiboítopí	pi	-ohoaó	hoíhio	piiohoaoxio.	
slowly	cut	river	-beside	larger	quantity by the river.	

Everett translates this sentence as, »Another day an old man slowly butchered a couple of big tapirs, by the side of the water.« However, he also reports that native speakers will reject as ungrammatical any addition to this sentence of a further adjective, such as would occur if one were to try to describe the tapirs in question as big and brown (Everett 2012, 558). Similarly, one cannot say in Pirahã, ›many big dirty Brazil nuts.‹ Instead one must say something like, ›There are big Brazil nuts. There are many. They are dirty‹ (ibid., 560).

More generally, Everett reports that Pirahã admits of sentences consisting of a verb's lexical frame plus at most one modifier word for each constituent of that phrase, and at most one prepositional adjunct phrase. Any proffered sentence with more complexity than this will be judged by Pirahã speakers as being ungrammatical (Everett 2012, 558). If, however, Pirahã were recursive, then no such upper bound would be possible. In a similar vein, Everett points out that Pirahã lacks disjunction, conjunction, and complement clauses. As such, Everett offers Pirahã as a counterexample to one of the central tenets of UG, namely that all normal speakers are competent in at least one language that is recursive.[9]

Proponents of UG might reply by invoking a distinction between what is sometimes called the faculty of language in the narrow sense, and the faculty of language in the wide sense, which correlates with the better-known distinction between I-language and E-language, respectively (Hauser *et al.* 2002). With

---

9 Gil 1994 also reports that Riau Indonesian lacks the property of recursive syntax.

the aid of this distinction, proponents of UG might respond to Everett's evidence from Pirahã by remarking that the most it shows is that the E-language is not recursive; but that leaves open the possibility that the I-language is. If, more precisely, the I-language is the seat of linguistic competence, while the E-language is the locus of performance, then proponents of UG might concede that Everett's data shows a performance limitation among speakers of Pirahã without needing to concede anything about their competence.

To this Everett could reply that the aforementioned pattern of Pirahã speakers' grammaticality judgments does not reflect a performance limitation. Rather, it is principled, and due to the fact that all information in a sentence must be grammatically marked evidentially, either as hearsay, as deduction from direct evidence, or as immediately experienced. (Everett (2012, 561) calls this the Immediacy of Experience Principle.) Everett argues that sentences whose grammatical complexity goes beyond the bounds specified above will necessarily contain information that is not marked evidentially, and so will violate the Immediacy of Experience Principle.

As should now be clear, the issues at stake are too deep to make it possible to determine the truth of UG here, or, most likely, any time soon.[10] However, the pattern that does emerge is that if UG turns out to be a scientific discovery, then this will not be due, even in part, to its being accepted by linguists and other members of the cognitive science community. Such acceptance will at most be evidence that UG is a scientific discovery rather than something that helps to constitute its being one. All that is necessary for UG to be a scientific discovery is that it be internally consistent, explanatory of a wide range of data, not refuted by other data, and apt to contribute to scientific knowledge. If these conditions are met, then UG will be *acceptable* to the appropriate cognoscenti, and this condition may be fleshed out in counterfactual terms as adumbrated above.

Accordingly, if the standard history of the »cognitive revolution« is correct, then Chomsky and colleagues discovered a fact about the human mind that was there and doing its work before anyone became aware of it. If, on the other hand, the challenges from Everett and others to UG prove cogent,

---

10   If I were to venture a suggestion concerning this debate, it would start with the observation that grammatically judgments should not be taken at face value. Most English speakers will likely deem ›Buffalo buffalo buffalo buffalo‹ to be ungrammatical, though we know that this sentence is grammatical. Likewise, most English speakers will likely judge ›He saw her‹, as ungrammatical if the two NP's in that sentence are understood as co-referring, but brief reflection shows that so understood, that sentence could be used to say something true. In this light, we may wonder whether Pirahã speakers' judgments of ungrammaticality are in fact judgments of another kind of impropriety.

then that will justify the conclusion that there is no species-wide language faculty as described with the three tenets of UG outlined above. (There may still be a species-wide language faculty that is not mandatorily recursive.) If however UG is correct, then we may use it to explain various phenomena, such as grammaticality judgments of ordinary speakers, as well as the surprising speed with which young children may learn a language. Likewise, we may contemplate a counterfactual scenario in which Chomsky and his followers had their work appropriated by other scholars who claimed credit, and then were given credit, for their discoveries. Here we would apparently want to say that although Chomsky and colleagues made a scientific discovery, the fact was not acknowledged by the scientific community.

These points about the status, both actual and counterfactual, of UG, conform to our general picture of the extent to which scientific discovery is a social phenomenon. For UG is a hypothesis which, if generally accepted, would add to what is impersonally known in cognitive science. But it can have that status, and thus be a scientific discovery, without changing what is impersonally known.

8.3.2 *Classifier Languages and Perception*

Many readers will be familiar with the Sapir/Whorf hypothesis according to which the language we speak affects our perception of the world (Sapir 1929, Whorf [1940] 1956). Whorf for example writes,

> Formulation of ideas is not an independent process, strictly rational in the old sense, but is part of a particular grammar, and differs, from slightly to greatly, between different grammars. We dissect nature along lines laid down by our native languages. The categories and types that we isolate from the world of phenomena we do not find there because they stare every observer in the face; on the contrary, the world is presented in a kaleidoscopic flux of impression which has to be organized by our minds – and this means largely by the linguistic systems in our minds. (Whorf [1940] 1956, 212, as quoted by Reines and Prinz 2009)

Whorf's line of thought stands in tension with a traditional Lockean view according to which language expresses and communicates thoughts that are fully formed independently of their linguistic expression. In the first decades after the popularization of the Sapir/Whorf hypothesis, its tendency to generate excitement tended to outrun its empirical credentials.[11] However, in recent decades researchers have found modest support for it.

---

11   Popular discussions of the Sapir/Whorf hypothesis have, at least in North America, often been carried on with reference to the allegedly large number of words for snow found in Inuit languages as compared with those found in English. See Pullum (1989) for an

One source of support concerns the difference between so-called classifier languages and others. English has both count nouns, such as ›dog‹ and ›chair‹, and mass nouns, such as ›water‹ and ›snow‹. The difference is that count nouns semantically ›unitize‹ their referents while mass nouns require something additional to bundle them into countable units. Mass nouns are bundled or unitized by *number classifiers*. For example, when you want to say how much water you drink you cannot say, ›I drink two waters a day‹. To count the referents of mass nouns for the purpose of saying that you drank two waters, you have to add a number classifier, such as ›glasses‹ or ›ounces‹ or ›gallons‹. By contrast, count nouns do not require a classifier. If we want to say how many cats are on the desk we can say, ›There are two cats on the desk‹. It is part of the semantic content of ›cat‹ that its referents are discrete entities. They are already in units and are therefore countable.

In some languages, known as classifier languages, all nouns require classifiers. They include Chinese, Japanese, and Yucatec Maya, a Mayan language spoken by some Native Americans of the Yucatan Peninsula (Lucy & Gaskins 2001, Lucy & Gaskins 2003). For English speakers, it is hard to conceptualize how classifier languages work. In our language, some nouns have a unit already built in as part of their meaning, like ›comb‹, ›book‹, or ›desk‹, and some do not, like ›water‹ and ›mud‹. In Yucatec Maya all nouns are like the English ›water‹ or ›mud‹. The result is that Yucatec Maya contains no terms that break up their referent into units or items. Even terms for ›book‹, ›cat‹, and ›table‹ require a classifier that specifies the unit. This suggests that Yucatec Maya nouns behave like stuff names. The classifiers specify how different kinds of stuff get put into units. For instance, to say ›two candles‹ in Yucatec Maya you have to say something like ›two long thin wax‹ where ›long thin‹ is the classifier.

In one of his early studies, Lucy (1992) showed that these linguistic differences affect memory. When presented with pictures of scenes that differ in either the amount of some stuff (say a pile of grain) or the number of objects (say the number of dogs), English speakers find changes of the latter sort easier to detect and recall. If a picture with four dogs is replaced by a picture with three, English speakers notice that, but might fail to notice changes in piles of grain. Yucatec Maya speakers do not show this pattern and are less likely than English speakers to notice when the number of dogs changes. Count nouns seem to encourage counting, and counting influences what we notice and recall.

---

amusing survey of how such claims have come to be made out of proportion to the available evidence.

In more recent work, Lucy and Gaskins (2001, 2003) argue that mass nouns draw attention to material substances, whereas count nouns draw attention to shape. These researchers presented English and Yucatec Maya speakers with three objects and asked them to judge which two of the three items were more similar. The triads were made up of a ›pivot‹ item, a material alternate, and a shape alternate. For example, one triad included a plastic comb with a handle as pivot, a plastic comb without a handle as material alternate, and a wooden comb with a handle as shape alternate. Another triad included a square piece of paper as pivot, a paper book as material alternate, and a square piece of burlap as shape alternate. Consistent with their prediction, English-speaking subjects chose the material alternate as the more similar only 23% of the time, whereas the Yucatec Maya speakers chose the material alternate 61% of the time.

These results suggest that dominant noun type in one's native language makes a difference for what you notice, co-classify, and recall. The nouns of a classifier language function more like stuff names, drawing attention to composition, whereas count nouns, when available, draw attention to shape and number. The nouns we use seem to make a difference in thinking.

As in the case of biological discovery, the discovery of the effects of classifier versus non-classifier languages is one that Lucy and colleagues had the conceptual and communicative skills to test, find evidence for, and then write up and publish. By virtue of those publications in refereed venues, the scientific community in effect validated the effect that Lucy *et al.* isolated and then described. In a pattern consistent with what we encountered in our discussion of Universal Grammar, this validation is optimal, but is not strictly necessary for these results to be scientific discoveries.

8.3.3   *Horn Scales*

Suppose a speaker S makes a claim C that is less informative than another claim C' that she might have made, where both C and C' would be relevant to the conversation to which S has contributed, and where C' is not considerably greater in length or syntactic complexity than C. Then S may in some cases be understood as meaning that she is not in a position to claim C'. John Stuart Mill (1865) noted this point,[12] but it was most famously taken up a century after Mill

---

12   »If I say to any one, ›I saw some of your children to-day,‹ he might be justified in inferring
      that I did not see them all, not because the words mean it, but because, if I had seen them
      all, it is most likely that I should have said so: though even this cannot be presumed unless
      it is presupposed that I must have known whether the children I saw were all or not.«
      (Mill 1864, 442)

by H.P. Grice, who hypothesizes that everyday conversation is governed by a set of usually unspoken norms that he calls conversational maxims. These maxims fall under a more general norm that he calls the Cooperative Principle.[13] On the hypothesis that speakers carry on their talk-exchanges in ways guided by the CP and the conversational maxims, Grice offers an explanation of the phenomenon mentioned by Mill. Crucial to that explanation is the hypothesis that the best explanation of the speaker's having spoken as they did is their desire to be cooperative while conforming to the Gricean maxims. That is why, says the theory, when a speaker reports, »Padma took the medicine and got better,« we normally infer that she meant that the medicine ingestion was prior to, and possibly the cause of, Padma's recovery. However, this implication is not a logical entailment due to the semantic content of the words used, as shown by the fact that the speaker may say, »Padma took the medicine and got better, though not necessarily in that order.«

Perhaps the bulk of research in linguistic pragmatics has focused on those implicatures that are generated by interaction with the maxim of Quality, which enjoins speakers to be as informative as, but no more informative than, is required for purposes of the conversation in which they are participating.[14] Levinson (1983) offers a line of reasoning that makes appeal to Quality, and that an observer of an utterance could in principle engage in to justify their imputation of a conversational implicature to a speaker:

> (i) S has said p.
> (ii) There is an expression q, more informative than p (and thus q entails p), which might be desirable as a contribution to the current purposes of the exchange,
> (iii) q is of roughly equal brevity to p; so S did not say p rather than q simply in order to be brief (i.e., to conform to the maxim of Manner)
> (iv) Since if S knew that q holds but nevertheless uttered p he would be in breach of the injunction to make his contribution as informative as is required, S must mean me, the addressee, to infer that S knows that q is not the case, or at least that he does not know that q is the case. (Levinson 1983, 106)

---

13   »Make your conversational contribution such as is required, at the stage at which it occurs, by the accepted purpose or direction of the talk exchange in which you are engaged.« (Grice 1989, 26)

14   »Make your contribution as informative as is required (for the current purposes of the exchange). Do not make your contribution more informative than is required.« (Grice 1989, 26)

In his 1972 UCLA dissertation, L. Horn (1972) argued that for many languages it is possible to construct sequences of words or expressions that are ordered in terms of logical strength, among the simplest cases in English being:

<some, many, most, all>
<ok, good, excellent, fantastic>
<possibly, probably, necessarily>

The notion of logical strength at issue may be defined as follows. Let ›φ[...]‹ be a sentence-frame taking terms as inputs to yield indicative sentences as outputs. Let t and t^{n+1} be terms both occurring on a single scale S. Then t^{n+1} is logically stronger than t just in case $\varphi[t^{n+1}]$ entails $\varphi[t]$ but not *vice versa*. A single term may occur in more than one scale.

On the assumption that speakers have an intuitive grasp of such scales as the above, we may now plug in any such scale to the above calculation from Levinson so that, for instance, when a speaker asserts that a particular book was good, she also means that she is not in a position to claim that the book was either excellent or fantastic.[15]

Subsequent literature in linguistic pragmatics has adopted the phrase ›Horn scales‹ to acknowledge Horn's work. However, it has been argued that step (iv) of the reasoning quoted above from Levinson misconstrues the maxim of Quantity as enjoining speakers to be as informative as possible so long as being more informative would not make them prolix or irrelevant (Green 1995). Quantity does not enjoin speakers to be as informative as possible, but only as informative as is required, and these two injunctions can come apart. To rectify the lacuna, any cogent reasoning along the lines of Levinson's reconstruction would have to refer not only to Quantity, but also to the notion of a question under discussion (as articulated in, e.g., Roberts 2012) or some other more detailed articulation of what a conversation requires such as that proposed by Green (2017), who delineates conversations into types.

Further, conversational implicatures are understood by Grice, and by most authors who address the matter, as being instances of speaker meaning, which is most commonly understood as involving intentions to produce a psychological effect on an addressee my means at least in part by the addressee's

---

15   Assume that such stronger claims would also be relevant to the conversation to which the speaker is contributing, and that the lexical difference between the adjectives in question would make a negligible difference for processing cost. Also, in certain cases we may impute to the speaker a stronger commitment, namely that she believes that the stronger claim is not true. (Geurts 2010 calls this the *competence assumption*.)

recognition of one's intention.[16] However, it has also been argued (Green 2019) that we do not need to impute such complex intentions to a speaker who is less informative than she might have been expected to be; instead we need only explain her diffidence on the hypothesis that she realizes that being more informative in such a situation would have resulted in her violation of Quality, which enjoins her to say only that for which she has sufficient evidence. One can be less informative than might have been expected without meaning that one can be no more informative, just as one might run less quickly than one has agreed to without meaning that one can run no faster.

Finally, Geurts (2010) argues that Horn scales are at best epiphenomenal, in that they can be fully explained by speakers' knowledge of the truth conditions of the sentences they produce and interpret. (We have already implicitly acknowledged this in our definition of Horn scales in terms of sentential frames.) Over and above our semantic knowledge of such sentences, there is no need to posit a further body of speaker knowledge of scales. If Geurts's criticism is correct, it would not show that Horn scales do not exist, but rather that positing them is superfluous relative to the posit that we already make in ascribing to speakers a grasp of the truth conditions of the sentences they grasp and the entailment relations among them. However, even if the posit of Horn scales is vindicated as against Geurts's criticism, this would not be due, even in part, to its being accepted by the community of professional linguists. Rather, it would be due to that posit's being independently motivated, perhaps on ground of its psychological reality, over and above the posit of speakers' semantic knowledge.

8.4   Conclusion

In this essay I have undertaken to examine the credentials of the view that scientific discovery must be understood from the perspective of institutional reality. Based both on general considerations about the possibility of unacknowledged discoveries, and on examination of three cases of putative discoveries in linguistics, I have argued that scientific discovery does not need to be understood through the lens of institutional reality. That negative result is compatible with the fact that the notion of scientific discovery, unlike personal

---

16   See Green 2020 for a discussion of speaker meaning. See Green 2019 for references to a variety of authors who take conversational implicature to be a species of speaker meaning.

discovery, must be understood with reference to what is known, and thus with impersonal knowledge.[17]

References

Everett, D.L. 2012. »What Does Pirahã Grammar Have to Teach Us About the Human Mind?« *Wiley Interdisciplinary Reviews (WIREs): Cognitive Science* 3 (6): 555–563.

Geurts, B. 2010. *Quantity Implicatures*. Cambridge: Cambridge University Press.

Gil, D. 1994. »The Structure of Riau Indonesian.« *Nordic Journal of Linguistics* 17 (2): 179–200.

Green, M. 1995. »Quantity, Volubility, and Some Varieties of Discourse.« *Linguistics and Philosophy* 18: 83–112.

Green, M. 2017. »Conversation and Common Ground.« *Philosophical Studies* 174: 1587–1604.

Green, M. 2019. »Assertion, Implicature, and Speaker Meaning.« *Rivista Italiana di Filosofia del Linguaggio* 13 (1): 100–115.

Green, M. 2020. *The Philosophy of Language*. Oxford: Oxford University Press.

Grice, H.P. 1957. »Meaning.« *The Philosophical Review* 66 (3): 377–388.

Grice, H.P. 1989. *Studies in the Way of Words*. Cambridge: Harvard University Press.

Hauser, M.D., N. Chomsky, & Fitch, W.T. 2002. »The Faculty of Language: What Is It, Who Has It, and How Did It Evolve?« *Science* 298: 1569–1579.

Hirschberg, J. 1985. *A Theory of Scalar Implicature*. Ph.D. thesis, University of Pennsylvania.

Horn, L. 1972. *On the Semantic Properties of the Logical Operators in English*. Ph.D. thesis, UCLA.

Horn, L. 2009. »WJ-40: Implicature, Truth, and Meaning,« *International Review of Pragmatics* 1 (1): 3–34.

Hyman, J. 1999. »How Knowledge Works.« *The Philosophical Quarterly* 49 (197): 433–451.

Levinson, S.C. 1983. *Pragmatics*. Cambridge: Cambridge University Press.

Lucy, J.A. 1992. *Grammatical Categories and Cognition: A Case Study of the Linguistic Relativity Hypothesis*. Cambridge: Cambridge University Press.

---

17   Michel (2019) also gestures at, without quite espousing, the claim that it is an essential feature of science that it can make discoveries. This claim would need further refinement before it can be adequately assessed. That refinement would I hope take into account that we seem to have a fairly clear concept of a completed science, in which all relevant types of phenomena within a certain domain have been explained. In that situation, I take it, there would be nothing left to discover in that domain beyond the etiology of particular events.

Lucy, J.A. & Gaskins, S. 2001. »Grammatical Categories and the Development of Classification Preferences: A Comparative Approach.« In *Language Acquisition and Conceptual Development*, edited by M. Bowerman & S.C. Levinson, 257–283. Cambridge: Cambridge University Press.

Lucy, J.A. & Gaskins, S. 2003. »Interaction of Language Type and Referent Type in the Development of Nonverbal Classification References.« In *Language in Mind: Advances in the Study of Language and Cognition*, edited by D. Gentner & S. Goldin-Meadow, 465–492. Cambridge: MIT Press.

Michel, J.G. 2019. »How Are Species Discovered? Declarative Speech Acts in Biology.« *Grazer Philosophische Studien* 96 (3): 419–441.

Mill, J.S. 1865. *An Examination of Sir William Hamilton's Philosophy and of the Principal Philosophical Questions Discussed in His Writings*. 2nd Ed. London: Longman, Green, Longman, Roberts & Green.

Pullum, G. 1989. »The Great Eskimo Vocabulary Hoax.« *Natural Language and Linguistic Theory* 7: 275–281.

Reines, M.F. & Prinz, J. 2009. »Reviving Whorf: The Return of Linguistic Relativity.« *Philosophy Compass* 4 (6): 1022–1032.

Roberts, C. 2012. »Information Structure in Discourse: Towards an Integrated Formal Theory of Pragmatics.« *Semantics and Pragmatics* 5: 1–69.

Russell, M. 2003. *Piltdown Man: The Secret Life of Charles Dawson and the World's Greatest Archaeological Hoax*. Stroud: Tempus.

Sapir, E. 1929. »The Status of Linguistics as a Science.« *Language* 5 (4): 207–214.

Schönfeld, M. 2006. »Kant's Early Cosmology.« In *A Companion to Kant*, edited by G. Bird, 47–62. Malden, Mass.: Blackwell.

Searle, J.R. 1989. »How Performatives Work.« *Linguistics and Philosophy* 12 (5): 535–558.

Searle, J.R. 1997. *The Construction of Social Reality*. New York: The Free Press.

Searle, J.R. 2007. »What Is Language: Some Preliminary Remarks.« In *John Searle's Philosophy of Language: Force, Meaning and Mind*, edited by S.L. Tsohatzidis, 15–45. Cambridge: Cambridge University Press.

Searle, J.R. 2010. *Making the Social World: The Structure of Human Civilization*. Oxford: Oxford University Press.

Whorf, B.L. [1940] 1956. »Science and Linguistics.« In *Language, Thought, and Reality: Selected Writing of Benjamin Lee Whorf*, edited by J. Carrol, 207–219. Cambridge, Mass.: MIT Press.

Williams, B. 1972. »Knowledge and Reasons.« In *Problems in the Theory of Knowledge*, edited by G.H. von Wright, 1–11. The Hague: Nijhoff.

The Discovery of Anthropogenic Climate Change

Jörg Phil Friedrich

Within the scientific community, there is a wide consensus that in our days there is an *anthropogenic climate change*. Also, there is a consensus that decades ago there was no process of this kind. Consequently, at some point in time somebody must have discovered this specific climate change, and the community must have accepted this finding as a discovery. In this paper, I will discuss the discovery of the *anthropogenic climate change* from a philosophy of science perspective. I will argue, that it is on the one hand a very special kind of a scientific discovery, but on the other hand, as I will show, this process of discovery sheds light on other kinds of scientific discoveries in theory, experimental and other empirical settings, and computer simulations.

9.1 Some Reflections on the Notions of *Change* and *Climate*

If we use the term *change*, we must distinguish between two possible meanings. First there is a gradual change in many objects, there is a shift, a slight change in the properties of objects everywhere. People grow and look older from one Christmas meeting to the other. The colours of my clothes change gradually during the years of use. Usually it is easy to deal with such slow changes. Although we do not mention these changes from day to day, we know that we need to handle them: We need to speak to a young adult in a different way than we spoke to the same person years ago when she was a child. We also know that we need to check the state of our clothing which we want to wear in the next season again, and it can happen that we need to buy some new items.

It is necessary to distinguish another kind of change from these slow and gradual shifts. Some changes are unexpected, fast, radical, and abrupt. Maybe I meet a good friend after a short time and her political position has changed radically. Maybe, my red clothes come out grey after the washing. In such cases I am surprised, it is not the normal change that I would have expected, it is, as we say, a »real change«.

When we use the notion of *change* in what follows, we speak about such abrupt, radical, »real« changes. It is important to state this clear because some people deny the importance of radical changes by claiming that there is change always and everywhere. And indeed, there was climate change in the sense of slow shifts in what we can call normal atmospheric conditions in every phase of Earth history. Furthermore, starting with the appearance of mankind there was an anthropogenic climate change. Ever since men started to burn down trees and sow grain, they altered parts of the climate system and so changed climate. If we use the term »climate change« today, however, we speak about a process that has never been there before. Anthropogenic climate change is a change in the Earth's climate system in a short time that has never happened before.

In addition, we need to reflect on the notion of *climate*, too. Today, especially in political discussions about anthropogenic climate change, the term »climate« is often understood as a global thing that can be described by a *global mean temperature* of the lower atmosphere, and climate change is often presented by the change of this temperature over time. However, this is neither the ordinary meaning of the word climate nor its scientific definition. For a proper understanding, it is helpful to start with the meaning of the term »climate« in its ordinary use, also in contexts beyond the weather conditions and the atmosphere, for example when we speak about a *business climate* or a *work climate*.

Climate is what you can expect in situations that you are familiar with. Think about the winter in the region where you grew up. After some years you knew what kind of clothes you need. In regions with a lot of snow, people usually learn to ski when they are children, while in other regions they do not.

You also know that there can be certain deviations from your expectations, no winter is exactly »normal«. You also know that there can be extreme exceptions to your expectations, but only every few years. For normal conditions with certain deviations, you have the right clothes and the right heating in properly insulated houses. Exceptions are dangerous, but because they are rare, you and the society can recover from exceptional weather events. The sum of theses experiences regarding the weather events in this region, the normal, the deviations, and the exceptions, are the climate.

Modern science put everyday experiences into a statistical description. Meteorologists measure the temperature, the amounts of rainfall, the wind speed and other attributes of the weather, and climatologists calculate statistical figures over certain periods of time. Consequently, for a long time, climate was a local thing. Until today, we can find this meaning in the definition published by the IPCC:

> »Climate in a narrow sense is usually defined as the ›average weather,‹ or more rigorously, as the statistical description in terms of the mean and variability of relevant quantities over a period ranging from months to thousands or millions

> of years. The classical period is 30 years, as defined by the World Meteorological Organization (WMO). These quantities are most often surface variables such as temperature, precipitation, and wind. Climate in a wider sense is the state, including a statistical description, of the climate system.« (Field et al. 2012, 557)

We are now in a better position to ask what the discovery of *anthropogenic climate change* amounts to. If we are following the scientific definition of climate, then a climate change is a significant shift in the figures calculated for describing the statistics. But then, two problems arise: first, we need to wait for 30 years to detect the climate change. Second and more important, we cannot be sure that the climate change is anthropogenic. In fact, we spoke about the reality of anthropogenic climate change with certainty long before we had a significant shift in the 30 years statistics, and we are sure, that the change is both dramatic and anthropogenic. How is that possible?

Let us start with the following assumption: If there was a time without an abrupt and dramatic anthropogenic climate change and we are now sure that there is such a process today, then we must have discovered this process at a certain point in time. This raises two questions:

1. When has this climate change been discovered?
2. Where was it discovered?

As to the first question: The discovery of climate change, as an event, has clearly happened at some point in history. But what kind of scientific event is it? An answer to this question could illuminate scientific discoveries in general, since the general question arises in what kinds of scientific work discoveries take place.

To find the moment in history where anthropogenic climate change was discovered, it should be helpful to have a look at the history of climate science, especially at the scientific events connected with the processes that we nowadays count as important for the understanding of climate change. On the following pages, we will discuss some major steps of this history and we will see in which way they have contributed to the discovery of the anthropogenic climate change.

9.2 The Discovery as an Effect in an Experiment

It is often said that almost 200 years ago, Jean Baptiste Joseph Fourier first described the heating of the atmosphere in a model that compared the Earth system to a *greenhouse* – a term we use until today to name the effect of global warming. It has been shown, however, that Fourier was not the first who used this picture, but that, as far as we know today, Edme Mariotte was the first to use it in 1681. Anyway, the idea of this picture is that the atmosphere (or parts of it) has the effect of a roof or a blanket that captures the heat that is reflected from the surface of the earth. The French physicist Claude Pouillet speculated in 1838 that water vapor and carbon dioxide could be the main greenhouse gases in the atmosphere (van der Veen 2000).

In the mid-19th century, experimenters discovered that different gas mixtures react differently to solar radiation with regard to their temperature change. They experimented with glass tubes filled with gases, but they inferred the consequences for Earth's atmosphere. In 1856, the American scientist Eunice Newton Foote derived not only the atmospheric greenhouse effect from her experiments, but also the consequences of changes in the concentration of carbon dioxide in the atmosphere:

> »The highest effect of the sun's rays I have found to be in carbonic acid gas. [...] An atmosphere of that gas would give to our Earth a high temperature; and if as some suppose, at one period of its history the air had mixed with it a larger proportion than at present, an increased temperature [...] must have necessarily resulted.« (Foote 1856, 383)

Imagine that one had asked her what would happen if mankind doubled the concentration of carbon dioxide within some decades. She would have replied that the temperature of the atmosphere would inevitably increase.

Consequently, we can imagine that a scientific discovery may be an event during the experimental work. We use the term *effect* here in accordance with the contemporary use for what Peirce called a *phenomenon*:

> »When an experimentalist speaks of a *phenomenon*, such as ›Hall's phenomenon‹, ›Zeemann's phenomenon‹ and its modification, ›Michelson's phenomenon‹, or ›the chess-board phenomenon‹, he does not mean any particular event that did happen to somebody in the dead past, but what *surely will* happen to everybody in the living future who shall fulfill certain conditions.« (Peirce 1905, 173)

As Habermas wrote: »The effects achieved under experimental conditions are each achieved in a *singular* experiment and nevertheless mean the assertion of a *universal* relation.«[1] Some effects only take place in experimental or technical settings, such as the Hall effect or the laser effect (Hacking 2010,

---

1 Habermas 1973, 163, my translation. »Die unter experimentellen Bedingungen erzielten Effekte werden jeweils in einem *singulären* Versuch erzielt und bedeuten gleichwohl die Feststellung einer *universellen* Beziehung.«

226). The conditions under which these effects occur are so special that it is almost impossible for them to occur under natural conditions and – even if they were realized somewhere in the universe – nobody would notice. So, for the scientific discovery of an effect scientists usually need a laboratory. Under controlled technical conditions, they can produce and reproduce effects, they can fix and describe the conditions under which the effect will occur and will be observable. Hence, we are well-justified in claiming that the one who first produced (and reproduced) an effect in an experiment is the discoverer of this effect.

A discovery of an effect in a scientific experiment meets all the characteristics of a scientific discovery described by Michel (2019). We will illustrate this for the case of Footes discovery, but it is obvious that the following remarks are valid for the production of effects in experiments in general.

First, for making a discovery of an effect in an experimental environment, »*a prior body of* [...] *knowledge*« (Michel 2019, 429, my emphasis) is necessary. This knowledge is twofold: on the one hand, the discoverer needs to have technical knowledge regarding the instruments in her laboratory. Foote, for instance, had to know that she could indeed produce carbon dioxide and that it really is in a certain concentration inside the glass tube. Moreover, she needed knowledge about the function of the thermometer, and so on. On the other hand, she needed to have some knowledge about the world, some ideas about the processes outside the laboratory. If she had no knowledge about real processes or no ideas about what could happen under certain conditions, she would certainly not have set up the experimental design.

Second, Footes discovery was *no pure coincidence*, and *third*, it was a *structured* discovery (cf. Michel 2019, 429) . Foote had a plan to prepare the glass tubes with different gas mixtures and she systematically varied the concentration of carbon dioxide and systematically measured the temperature inside the tubes.

Fourth, the discovery was *accepted* by the scientific community (cf. Michel 2019, 429) – and it is an interesting detail that, at first, another scientist carrying out nearly the same experiment was attributed to be the discoverer of the effect. *Fifth*, the importance of *clear credits* (cf. Michel 2019, 429f.) for scientific discoveries can be seen in this case quite well because the discussion about the question who really was the first to produce, observe and describe the effect affected the scientific community after decades anew.

Finally, *sixth*, the case also shows the importance of *language* for a scientific discovery (Michel 2019, 430): It is the description of the discovered effect in a scientific publication that makes it possible that the discovery is accepted and that we can attribute the credits of the discovery to Foote.

However, the discovery of the effect of warming of air mixed with a high portion of carbon dioxide is obviously not the same as the discovery of anthropogenic climate change. But, why not? We could say that the effect Foote discovered is exactly the same as the effect of global warming today, and we could also say that anthropogenic global warming started during the 19th century, i.e., around the time when Foote carried out her experiments. On the other hand, the anthropogenic climate change is a process out there in the atmosphere of the Earth, therefore, we must say, the discovery should be made out there and not, as in Foote's case, in a laboratory.

9.3   Surprising Consequences of a Theory

Before we consider the question how to find a process like climate change out there in Earth's atmosphere, we will consider some other places of investigation. Around the beginning of the 20th century, Svante Arrhenius explored the effect of greenhouse gases in the atmosphere by theoretical means. Based on Stefan's Radiation Law, Arrhenius build the first theoretical model of the atmosphere already in 1896, and he calculated the influence of the atmospheric carbon dioxide concentration for the global temperature of Earth's atmosphere (Arrhenius 1896). He calculated that the atmosphere would be about 6 Kelvin warmer if the concentration of carbon dioxide would set to double.

In 1908, he published his finding that the burning of coal had the consequence of warming the atmosphere. Arrhenius, however, regarded this a positive consequence of the era of industrialisation. From his point of view, it would prevent mankind from a new ice age. Furthermore, by

> »the influence of the increasing percentage of carbonic acid in the atmosphere, we may hope to enjoy ages with more equable and better climates, especially as regards the colder regions of the earth, ages when the earth will bring forth much more abundant crops than at present, for the benefit of rapidly propagating mankind.« (Arrhenius 1908, 63)

Arrhenius' calculations were challenged and corrected in detail by other scientists during the first decades of the 20th century,[2] but in the end the scientific community accepted them in principle. In the light of this, we can say, that the theoretical case of the discovery of climate change also meets the aforementioned characteristics of Michel. Before we discuss this in detail, however,

---

2   For a summary, see Fleming 2007, 69.

we can derive some insights regarding discoveries in the theoretical part of a scientific discipline from this case.

Within a scientific theory, or within the theoretical part of science, a discovery can be seen as a (surprising) *consequence of a theoretical model* under certain conditions. Consider the case of theoretical physics using mathematical equations as models. In this case, a scientist can make some assumptions for filling some terms in the equations and she can use some values for the variables in the equations for stating the starting and the boundary conditions for a special situation she wants to describe with the model. With these steps of preparation, the theorist can solve the equations and, consequently, calculate the development of the values of the variables under the defined conditions. If she finds an unexpected solution, or even a qualitatively new solution of the model, we can call this a *theoretical discovery*.

Theorists do not need to care about reality. They could, for instance, calculate the atmospheric conditions for a hypothetical planet in a fictional part of the universe and could find a surprising behaviour of the temperature on such a planet. Albeit its fictional character, this could be a theoretical discovery. By contrast, regarding a theoretical finding concerning Earth's atmosphere, the theoretical setting somehow has to be realistic. But, what does that mean?

Challenging a theoretical discovery is possible in two different ways. First, we can try to find an empirical setting meeting the theoretical starting and boundary conditions of the theoretical model. This is possible by performing experiments in the laboratory, but also by measuring real processes out there in the field, in nature. This is what Popper had in mind when arguing for his falsificationist view. Strictly speaking, however, it is not the theory that is falsified by empirical investigations, but a theoretical model based on a theory and on some realistic assumptions fitting the theoretical model to reality out there.

Regarding the case of *anthropogenic climate change*, for Arrhenius and his colleagues it was not possible to challenge or to prove his theoretical finding in this way. They thought to measure the effect would be possible only in the far future. In fact, as we will see in the next section, only a few decades later, Callendar tried to find the anthropogenic climate change in meteorological measurements of atmosphere's temperature.

Nevertheless, there is another way to challenge a theoretical finding, and by doing so, determining whether it is indeed a scientific discovery with respect to reality: First, one can check if the calculations are correct. Second, one can discuss if the theoretical model is derived properly from the theories and if all assumptions made are acceptable. Third, one can check if the starting and boundary conditions in the model are realistic, that means, if scientists are able to imagine a path or a scenario for reality to come from the present state of the real world to the state the model starts with. Fourth, the scientific community may discuss other processes in nature, not included in the model, but affecting the development of the process that is theoretically described.

To make a long story short: All these possibilities of challenging Arrhenius' finding were carried out during the first decades of the 20th century, and the result was that, in the end, the scientific community accepted his theoretical findings. It needs to be noted that his finding would still be correct even if humans stopped burning coal and oil, and even if other processes like air pollution or the breakdown of sun radiation lowered atmospheric temperatures.

We can see that a theoretical discovery meets all the attributes of a scientific discovery mentioned by Michel as well, as can easily be seen with the help of this example. First, a scientist needs to have prior knowledge: Arrhenius had to know the theories, the proper way to derive and manipulate models from it, and the information about realistic or interesting starting and boundary conditions. Second, the theoretical discovery is no pure coincidence, it was guided by a theoretical question and by the experimental results mentioned before. Third, it was a structured discovery, because the building and manipulation as well as the calculation of the results is a structured process guided by the rules of the theoretical discipline.

Fourth, the discovery is finally accepted by the scientific community. At first, Arrhenius' finding was questioned by other scientists, it was discussed and finally accepted as a new discovery. Fifth, we can clearly credit Svante Arrhenius for the discovery. Sixth, the importance of language for a scientific discovery is obvious in the theoretical case: Constructing and manipulating a theoretical model is done by using the language of theory, in our case the language of mathematics, and the results are communicated in this language, too.

9.4   Finding a Signal in Statistical Data

We have already seen that for the discovery of anthropogenic climate change neither the discovery of an experimental effect nor the discovery of a surprising consequence of a theoretical model would be enough. The reason is, that the thing to be discovered is neither something we can produce at the laboratory bench nor is it a theoretical construction. It is, if it exists, something ›out there‹, in nature, in the real atmosphere, outside of all experimental settings and theoretical calculations. Consequently, we need to make the discovery there. But to do this, we need to solve two problems first. We need to define clear criteria to determine that the thing *is really* out there, and we need to make sure that the thing out there is what we think it is. That means that, first, we have to define what a climate change *is* and how we can be sure, that *there is* a climate change. Second, we must make sure that the climate change we discovered is an *anthropogenic one.*

As mentioned before, for climate scientists, climate is a statistical thing, and so, climate change is a significant change in the statistical description of the atmosphere. First, the climate was described by the statistical distribution of the values of the variables describing the weather of a place or a region. In the middle of the 19th century, meteorologists started to measure the attributes describing the weather in a systematic way. As of this time, they measured temperature, precipitation, wind direction and velocity, degrees and kinds of cloudiness, air pressure, and the moisture of the air in certain geographic places every day at the same time, and they recorded these measurements in protocols. On this basis, they started to calculate statistics: the mean values of the years or of a season or a month, but also the mean values for each month over time. The time interval for the statistics for describing the climate of a given place was 30 years. Therefore, climate change can be defined in a significant change in the statistics from one 30-year interval to the next. By means of a 30 years statistics of, say, the mean temperature of January, one can determine whether this January was »normal« or exceptional (»too warm« or »too cold«).

The question arises, how many »too warm« Januaries or »too warm« years are needed to determine that there is a climate change. One possibility is to compare the statistics of two 30-year intervals and to calculate by means of mathematical statistics whether the differences in the statistics are significant or not. Another option is to use statistical methods of time series analysis to see if there is a tendency or a shift in the measured values over time.

The first scientist who searched for a signal of anthropogenic global warming – and who found one – was Guy Stewart Callendar (Callendar 1938). From 147 land-based meteorological stations, he calculated a global mean temperature and analysed the time series for over 50 years, and he found a significant increase of this temperature. Furthermore, Callendar saw a causal connection between this increase and the burning of coal and found that the increase of the temperature could be explained by the increase of the concentration of carbon dioxide rooted in the coal burning.

Callendar's results, however, were not accepted by the scientific community. Of course, the increase of his calculated mean temperature could be explained by other causes than the increase of the concentration of carbon dioxide due to the burning of coal.

Today scientists use the moving average over about 10 years, calculated especially for the global mean temperatures per year, to analyse time series of mean temperatures and to show the change in the climate. Visualisations of the development of these values in time show dramatically, that something has changed during the last 30 years. The so-called *hockey stick curve* is probably the best example.

We can say that the discovery of a process in empirical statistical data of a time series proves – like a measurement of the yearly mean temperature at a given place over years – that there is a clear significant shift in these data. In general, we call such a shift a *signal* in the statistical data. Usually, we can see such a signal clearly in plots of the development of the value in question over time. Even in papers in scientific journals, scientists depict diagrams with plots of time series of mean temperatures showing the signal of change. By pointing to these signals, they state that there is the evidence of the change.

Supposing that a statistical evaluation of measurements of physical quantities represents something real, we can say that finding a *signal of change* in these statistics is a discovery of something real. Let us use the hockey stick curve as an example of a signal of change. Obviously, a discovery of a signal of change in climate statistics meets the characteristics of a scientific discovery. As before, prior knowledge regarding mathematical statistics is required, but also regarding the definition of climate and regarding the measurement of meteorological data. The discovery of the hockey stick was far away from being pure coincidence: it was the result of systematic search in statistical data. In the scientific community, it is widely accepted that the signal of change as seen in the hockey stick exists. Michael E. Mann and his colleagues are clearly credited for it connected with their publication in 1999 (Mann et al. 1999). They described what they found using the language of mathematics.

Therefore, we can say there is a scientific discovery of a signal of change in statistical data regarding the climate. But, still, the question remains whether this discovery is the discovery of the anthropogenic climate change

9.5   The Discovery as the Right Interpretation

In the following, things will be getting a little more complicated. Until now, we have discussed discoveries as relatively simple stories of actions. People are searching for something in a special way and they found something – and because the way of searching was motivated and grounded by prior knowledge, because the search was systematic and the finding not pure coincidence, furthermore because the scientific community accepted the finding and attributed clear credits to the discoverers – the finding can be seen as a scientific discovery, may it be an experimentally shown effect, a surprising consequence of a theory or a signal in statistical data.

The *anthropogenic* climate change, however, is not simply a thing we may find, not even by scientific means, it is an *interpretation* of the finding as a *special thing*. The experimental effect is *interpreted* as a process in the atmosphere producing the climate change, the theoretical consequence is *understood* as a theoretical description and an explanation of the climate change. Now, together with the insights (a) that humans produce the conditions for the effect in the atmosphere, and (b) that the amount of carbon dioxide produced by humans fits the theoretical calculations, we can *interpret* the signal in the statistical data as the anthropogenic climate change. Finally, after this overview of experimental, theoretical, and statistical discoveries, we are justified in claiming that the discovery of the hockey stick curve is to be regarded as the discovery of anthropogenic climate change.

9.6   Discoveries in Computer Models

There is a fourth field of scientific endeavour strongly connected with the research regarding the anthropogenic influence of humans on the climate, namely the numerical description of the atmosphere using mathematical models calculated by supercomputers.

The use of computer models for modelling atmospheric processes started with the numeric weather prediction in the 1950s (Shuman 1989). Climate models rely on the same principle but have more components. Not only are they models of the atmospheric processes, but they are coupled models of atmospheric and oceanic processes, of the variation of the ice masses and of the vegetation systems on Earth. One can run simulations for all these processes modelling what may happen in these systems over decades – and with these simulations, one can calculate model values for the physical parameters and calculate a *climate* of the model as one can measure real physical parameters and can calculate a climate for them. To tell a long story in short, scientists can calculate changes in the statistical description of ensembles of computer simulations. This mathematical description we can interpret as the climate of the model for different scenarios of the growth of the concentration of carbon dioxide and can show how the climate, defined as the statistical description of parameters modelled, in the model changes.

One can interpret such a finding in computer model as a discovery of climate change if one accepts that the numeric model calculated by a supercomputer says something about the real world. Following Krohs (2008), we can interpret the numerical calculation in the computer as well as the one of the processes in the atmosphere, the ocean, the ice shield, and the vegetation as »real systems« – both described by the same mathematical equations. Then, the observation and description of one real system – in our case, the computer simulation – give insights about the other system which cannot be observed in the same way because the processes are to slow. With this, we can discover the climate change of the future Earth climate systems already today in the observation of the computer simulation.

Just like discoveries of effects in experiments, discoveries of surprising consequences in a theory, and discoveries of signals in statistical data, discoveries in computer models meet Michel's characteristics for scientific discoveries. Scientists constructing and investigating these models need prior knowledge to build the models, they work systematically, so that their findings are not coincidental. There is only one exception regarding Michel's criteria in the special case of numerical climate modelling, and this is the criterion of clear credits. Since climate modelling with supercomputers is not only a joint enterprise of scientists from different disciplines in big teams, there are lots of teams working on different models all over the world and putting their results together, discussing them and finding more and more that is helpful for a common understanding of *what* the anthropogenic climate change really is, *that* it really exists, and *how* it proceeds.

Now, what kind of discovery is the discovery of climate change in computer models? Surprisingly, it is each of the kinds we discussed in the previous sections. First, because the computer simulation is an experiment, the finding is an experimental *effect*. Second, because the simulation is a calculation of a theoretical model, it is a more or less a surprising *consequence* of the theory. Finally, because the analysis of the data produced by the simulation is a statistical one, the finding is a *signal* in the statistical data. This suggests a deeper structural correspondence between the different areas of scientific work, which is discussed in detail elsewhere as the construction and manipulation of models (Friedrich 2019, 84–104).

9.7   Summary

Upon examining the case of *anthropogenic climate change*, we have identified three different kinds of scientific discoveries: In experiments, scientists discover *effects*, in theories, they discover *surprising consequences*, and in empirical statistics, they discover *signals*. We have seen that discoveries of one kind may become part of the *previous knowledge* in discoveries of another kind. This is also the reason why these discoveries are typically no pure coincidences: they can guide the systematic quest for discoveries in other parts of the scientific realm. For instance, an effect discovered in an experiment can be the starting point for a systematic analysis of theoretical models searching for boundary and starting conditions to derive the effect as a consequence of the theory. Both the experimental effect and the theoretical consequence can initiate the search for signals in empirical data. Furthermore, we have seen that simulations with computer models show aspects of all the other kinds of scientific discoveries.

After all, we are in a position to answer the following question: Where and when was the anthropogenic climate change really discovered? The surprising consequence of our considerations is: This depends on what we mean by the term *anthropogenic climate change*. Is the term used to refer to the effect of warming air with higher concentration of carbon dioxide? This was discovered in the mid-19th century by Eunice Newton Foote. This would hold even if we take into account that Foote knew nothing about the possibility of a dramatic increase in the carbon dioxide concentration by burning coal. After her experimental findings, she thought about consequences of higher or lesser concentrations of carbon dioxide in the atmosphere for the temperature of the air. Her conclusions correctly describe what we call *global warming* today.

But for us today, anthropogenic climate change is more than just the warming of the atmosphere by higher concentrations of carbon dioxide. It is at least the consequence of the real burning of coal and oil by humans. Therefore, the discoverers of the anthropogenic climate change can only be scientists taking into account this real cause of increasing carbon dioxide concentrations. Against this background, we can regard Svante Arrhenius and Guy Stewart Callendar as the discoverers of the anthropogenic climate change. Even though their findings were challenged and also revised by other scientists afterwards, they found the truth about the consequences of burning coal and oil.

Arrhenius and Callendar conceived of the anthropogenic climate change as a slow shift of atmospheric conditions over centuries. As I stated at the outset, however, climate change, as we see it today is an abrupt, radical, and fast change of the whole climate system, including oceans, ice shields, vegetation, and society. It is a grand challenge for our civilisation because it is far from clear that and how we will be able to adapt to this change successfully in the near future. *Anthropogenic climate change*, as we see it today, was discovered during the very last decades, mostly with the help of computer simulations, as the statistical signal in the hockey stick curve, and to a certain extent in the weather events of the last few years.

Perhaps the discovery of real anthropogenic climate change is not a scientific one. It will be the experience of everyday challenges in our future.

Acknowledgements

I would like to thank Jan G. Michel for critically reviewing the manuscript, for the many critical tips, and for the stimulating discussions that contributed to the success of this article. I would also like to thank Joseph Moore for his encouraging and helpful hints and the participants of the Making Scientific Discoveries workshop in Münster for helpful discussions.

References

Arrhenius, S. 1896. »On the Influence of Carbonic Acid In the Air Upon the Temperature of the Ground.« *The London, Edinburgh, and Dublin Philosophical Magazine and Journal of Science* 41: 237–276.

Arrhenius, S. 1908. *Worlds in the Making: The Evolution of the Universe.* New York & London: Harper & Brothers.

Callendar, G.S. 1938. »The Artificial Production of Carbon Dioxide and Its Influence on Temperature.« *Quarterly Journal of the Royal Meteorological Society* 64 (275): 223–240.

Field, C.B., Barros, V., Stocker, T.F., Qin, D., Dokken, D.J., Ebi, K.L., Mastrandrea, M.D., Mach, K.J., Plattner, G.-K., Allen, S.K., Tignor, M., & P.M. Midgley, eds. 2012. *A Special Report of Working Groups I and II of the Intergovernmental Panel on Climate Change (IPCC).* Cambridge: Cambridge University Press.

Fleming, J.R. 2007. *The Callendar Effect.* Boston: American Meteorological Society.

Foote, E.N. 1856. »Circumstances Affecting the Heat of Sun's Rays.« *The American Journal of Science and Arts* 22 (66): 382–383.

Friedrich, J.P. 2019. *Ist Wissenschaft, was Wissen schafft?* Freiburg: Alber.

Habermas, J. 1973. *Erkenntnis und Interesse.* Frankfurt: Suhrkamp.

Hacking, I. 2010. *Representing and Intervening: Introductory Topics in the Philosophy of Natural Science.* Cambridge: Cambridge University Press.

Krohs, U. 2008. »How Digital Computer Simulations Explain Real-World Processes.« *International Studies in the Philosophy of Science* 22 (3): 277–292.

Mann, M.E., Bradley, R.S., & Hughes, M.K. 1999. »Northern Hemisphere Temperatures During the Past Millenium: Inferences, Uncertainties, and Limitations.« *Geophysical Research Letters* 26 (6): 759–762.

Michel, J.G. 2019. »How Are Species Discovered? Declarative Speech Acts in Biology.« *Grazer Philosophische Studien* 96 (3): 419–441.

Peirce, C.S. 1905. »What Pragmatism Is.« *The Monist* 15 (2): 161–181.

Shuman, F.G. 1989. »History of Numerical Weather Prediction at the National Meteorological Center.« *Weather and Forecasting* 4 (3): 286–296.

van der Veen, C.J. 2000. »Fourier and the ›Greenhouse Effect‹.« *Polar Geography* 24 (2): 132–152.

Discoveries in Environmental Studies?

Joseph G. Moore

10.1   What is Environmental Studies Exactly?

Environmental studies is an increasingly popular field of study in the United States and around the world. Not only do most institutions of higher education have undergraduate programs in environmental studies, but the field now has its own graduate programs, dedicated journals, and scholarly societies. It was only fifty years ago that the first programs in environmental studies were established, and now the field is exploding. Interest in environmental studies is entirely understandable since the field promises to address – to understand and to help rectify – some of the most pressing problems facing our planet: overpopulation, biodiversity loss, depletion of natural resources, food security, energy use, water supply, pollution, environmental injustice, and the mother of all environmental challenges: climate change. Today's younger generations grow up with an awareness of the existential environmental challenges that confront their future. They learn basic environmental facts and concepts in grade school, and they wrestle with the environmental dimensions of their actions in domestic, social and political spaces. And of course, environmental issues are now ever present in the cultural background, even during a global pandemic that (as I write) focuses attention onto our own safety, health and economic well-being. So, students come to university eager to learn more in an academic way about their imperiled environment, as well as to develop tools to confront environmental challenges with action. It's no wonder that environmental studies is an exploding field.

Still, one might ask: what exactly is environmental studies as an academic study? Here's one answer drawn from my own department's website:

> For many thousands of years, our ancestors were more shaped by the environment, than they were shapers of it. This began to change, first with hunting and then, roughly ten thousand years ago, with the beginnings of agriculture. Since then, humans have had a steadily increasing impact on the natural world. Environmental studies explores the complex interactions between humans and their environment, and how these might change. This exploration requires grounding in the natural sciences, the humanities, and the social sciences.[1]

---

1   Drawn, with slight modifications from: www.amherst.edu/academiclife/departments/ environmental_studies/about.

Notice three nice features of this characterization. First, it's topic-focused – the field is characterized in terms of what it studies, the environment. Fixing on a domain of investigation is a natural way of characterizing any interdisciplinary field. Consider how we understand the fields of American studies, architectural studies, black studies, European Studies, film & media studies, Latinx and Latin American studies, sexuality, women's & gender studies – just to name some interdisciplinary programs at my own institution.[2] Second, the characterization emphasizes how environmental studies engages inter-disciplinarity on steroids. Environmental studies includes certain investigations in the natural sciences (e.g., in biology, climatology, and geology), but also in the social sciences (e.g., in economics, political science, sociology) and in the humanities (e.g., in history, philosophy, religion, and literary studies).[3] As such, environmental studies is the only field I know of that spans the »two cultures« divide between the sciences and the humanities.[4] Finally, the characterization also hints at the field's normative dimension. Environmental studies seeks to understand not just how we, humans, in fact shape, and are shaped by our environment, but also how we *should* interact with our environment.

This topic-focused, interdisciplinary, and partly normative characterization works rather well. It draws students, attracts faculty, excites administrators and gives everyone a good enough sense of what goes on in the field. Nevertheless, one might still wonder whether there is anything to environmental studies as a field beyond the cobbling together of findings from its contributing disciplines – geology, biology, economics, political science, history, philosophy, and so on. Environmental studies might be wonderfully interdisciplinary and important, but, as an intellectual matter, is it held together by anything more than its topic? Does it have any unique methodology, principles, laws or standards of rigor beyond those it derives from its contributing disciplines? Is there any sense in which environmental studies is a »discipline« in its own right?

In this paper, I wrestle with this issue of disciplinarity. First, I'll say a bit about why we might care whether environmental studies counts as a discipline. Second, I'll suggest that the issue can be usefully sharpened by asking whether environmental studies generates new knowledge in the form of discoveries. Third, I'll pursue some examples – one at some length – that support a positive answer to this question. And finally, I'll suggest that these discoveries belong centrally to environmental studies because they are guided by an

---

2   Going beyond Amherst College, consider: cognitive studies, indigenous studies, forestry, …

3   It's common, at least in the United States, to distinguish between programs in environmental studies and programs more narrowly in environmental science. So, ›studies‹ as used in the label ›environmental studies‹ has the more inclusive sense of the German ›Wissenschaft‹ – any systematic investigation.

4   See Snow 1959.

implicit »eco-holism« which should serve, I propose, as the field's »disciplin-ary lens«.

10.2   Why Care about Disciplines?

Academic disciplines are like prisons. Or so claimed the French philosopher, Michel Foucault in his 1975 book, *Discipline and Punish: the Birth of the Prison*:

> The [academic] disciplines characterize, classify, specialize; they distribute along a scale, around a norm, hierarchize individuals in relation to one another and, if necessary, disqualify and invalidate. (Foucault 1995, 223)[5]

With Foucault's warning in mind, one might ask why a refreshingly interdisci-plinary field like environmental studies would even aspire to be a discipline. After all, the field promises to uncover »outside of the box« approaches to the environment of just the type that might be discouraged by the disciplinary bar-riers Foucault warns against.

Foucault points to negative features of disciplines considered as *institutional groupings*. To my mind, not all these features, for example, specialization, are clearly negative. More importantly though, while Foucault is surely right that institutional groupings can create unhelpful barriers, disciplines can also be highly efficient mechanisms of social epistemology: they enable, promote and support the type of conversation, dialectic and collegial investigation that is central to intellectual progress. So, *pace* Foucault, healthy academic disciplines needn't be prisons, but rather springboards to greater understanding.

In any case, as a category of institutional grouping, environmental studies already resembles traditional disciplines in many ways. There are departmen-tal meetings, talks, societies, Ph.D.'s and so on. So, the important question is why we should care whether environmental studies counts as a »discipline« in a less institutional, and more intellectual sense of the term. The answer, I think, is because of a felt need for intellectual legitimacy.

Higher education is torn between the pedagogical goal of teaching estab-lished knowledge and skills, and the more scholarly goal of, to put it very broadly, advancing human understanding – uncovering new knowledge, new investigative techniques, new technologies and new pathways to cre-ative achievement. A foundational assumption of modern universities is that these two goals can be happily aligned when its faculty teach what they do:

---

5   Foucault thought this in part because he saw our academic divisions and the modern penal
    system as arising together in the social and political context of 18th-century France.

when students are pulled into the mission of uncovering new knowledge, they acquire established skills and practical understanding as a sort of happy side-effect. Non-traditional inter-disciplinary fields of study – »mere fields« – can seem to threaten this alignment by sacrificing something intellectual.[6] And so, the skeptic might worry that environmental studies, if it turns out to be a mere field, doesn't advance the university's intellectual goal. After all, academic inquiry is meant to advance human understanding not by shedding any old light on a topic, but to do so in a distinctive manner that is careful, methodical, systematic, rigorous and, well, disciplined. Not anything goes: we wouldn't sanction a department of Harry Potter studies, even if it generated large donations and massive student interest. So, it's a desire for intellectual legitimacy that might give environmental studies disciplinary aspirations.

Everyone feels this intellectual concern, including students and administrators who one might think prioritize the pedagogical goal. But the concern is felt acutely by faculty, myself included, who strive to achieve expertise, status and fulfillment through hard work, sophistication and rigor that is not just pedagogical but intellectual. The concern can manifest in the fierce battle over resources (offices, budgets, appointments, promotions, etc.). But the root unease stems from a protective sense of honoring the intellectual goal of the academy.

This unease is also felt, in a quite different manner, by parents: studying the environment is all well and good, but beyond heightened awareness and concern, what types of rigorous (and marketable) knowledge, methods, or skills might my child get from spending several formative years taking grab-bag of courses about the environment? Training in environmental studies may provide breadth, but where's the intellectual depth?

Of course, environmental studies will still be in good standing even if it turns out not to be a discipline in any significant non-institutional sense. The field is just too popular, and important. Moreover, its status is buoyed by standing alongside many other well-respected inter-disciplinary fields of study: film studies, black studies, cognitive studies, womens & gender studies, and American studies, for example.[7] Nevertheless, the question of disciplinarity is worth pursuing: a positive answer would allay skeptics, while even a negative answer might reveal a considered, liberating way in which environmental studies can wear the label »mere field« as a badge of honor.

---

6   Moreover, fields like accounting and forestry sit uneasily with the even stricter conception of the liberal arts – i.e., intellectual inquiry that is in principle unfettered entirely from practical application.

7   If we still feel insecure, we might pursue a *tu quoque* against established disciplines: it's not clear on reflection that philosophy, economics or even geology can be given more than a topic-based characterization.

10.3   From Discipline to Discovery

Whether environmental studies is a discipline depends, of course, upon what we mean by an academic »discipline«. We've already encountered an institutional sense in which environmental studies comes close: like any traditional discipline, environmental studies is already an institutionally recognized way of grouping and organizing scholars so as to investigate a clear enough topic.[8] But we've also seen that the label ›discipline‹ cries out for something more. We often think of a discipline as bound by a shared methodology – experimental protocols, standards of proof, quantitative or qualitative analyses, archival techniques, textual strategies, standards of critical experience and so on. And so, when we think of economics or history, for example, we think not just of a specific topic, but of a distinctive *way* of investigating this topic. For some disciplines, most notably the natural sciences, we might also point to agreed-upon principles or revealed laws that are somehow foundational.

The laws, the principles or the methods that characterize a discipline can change over time, of course, and sometimes with revolutionary speed. Moreover, disciplines are broad, with rough boundaries. So the most we can say is that certain methods and principles are currently »central« to a discipline. Still, a shared methodology or set of principles is a sign that there's something intellectually distinct and unifying that binds together the intellectual activities that fall under a discipline.[9]

It's not clear that any of this can be said of environmental studies: is there a shared methodology or set of foundational laws or principles that binds the field? The answer is easy if we adopt an inclusive conception according to which a field inherits the features of its contributing disciplines. In this inclusive sense, yes: environmental studies deploys the methods and principles of

---

8   Soulé and Press propose a different institutionally-grounded notion of disciplinarity: disciplines have a pedagogical curriculum that is »vertically« structured, or based around fundamental works that are agreed upon (1998, 298 ff.). I won't pursue this proposal since we're in search here of a *non-institutional* notion of disciplinarity. (Additionally, I worry that their necessary condition implausibly rules out clear disciplines such as history.) Still Soulé and Press say many useful things about the structure of environmental studies curricula.

9   This holds, by the way, even of my own legacy discipline, philosophy. Philosophers won't universally agree to anything, especially the proper way to carry out their own study. Even so, if philosophers were polled about different characterizations of their discipline, I suspect there would be wide-spread agreement about what counts as philosophy and why. Here's one attempt: philosophy focuses distinctively on the adequacy of theories and arguments that aim to illuminate the more abstract features of human nature and the natural and social worlds we inhabit, even when – especially when – these theoretical ideas challenge the unexamined presuppositions and lacunae of everyday life.

biology, of economics, of history, of philosophy, and of all of its other contributing disciplines bundled together. But this bundle of methods and principles
is more a motley crew than a unifying lens. So, the question is properly asked
non-inclusively: does environmental studies have methods or principles *of its
own*, beyond those derived from its contributing disciplines? Asked in this way
the answer is no: it's hard to see that the field has any methods or principles
all of its own. In this way, environmental studies remains, at root, an interdisciplinary field.

But even if there are no principles or methods unique to environmental
studies, there's a close cousin that's worth exploring – *new knowledge*. Does
environmental studies uncover any knowledge that doesn't arise from its contributing disciplines? Does environmental studies offer explanations, understandings or even discoveries that are interestingly synthetic – that flow from
environmental studies holistically, and not just from cobbling together the
findings of its contributing disciplines?

I want to focus here on discovery, because it's a particularly dramatic and
tractable form of new knowledge. Not all explanations and understandings
count as discoveries, of course. The realm of art offers new ways of seeing or
understanding the world in the form of creations, not discoveries. Still, asking whether there can be discoveries that are unique to environmental studies
is, I think, a useful way to test whether environmental studies is importantly
discipline-like. I'll leave for later, the question of whether a field that gives rise
to unique discoveries should thereby count as a discipline. I'll turn now to
the question of whether there are even any candidate discoveries that belong
uniquely to environmental studies.

10.4   Some Discoveries in Environmental Studies

I've asked a few colleagues working in environmental studies what discoveries,
if any, they think have been made within the field. Here's a partial list, which
I'll present without commentary or much explanation:

> Technologies: solar panels, biofuels, genetically modified organisms.

> Causal connections: overfishing affects speciation (favoring small, quickly
> maturing fish); the reintroduction of wolves caused increased river meander
> ing in Yellowstone National Park; Columbus caused the Little Ice Age.

> Prominent concepts: ecosystem services, the tragedy of the commons, envi
> ronmental justice, anthropogenic climate change, disturbance regimes,

the Anthropocene, the IPAT/Kaya formulae,[10] demographic transition, eco-holism.

This fairly captures, I think, the range of things one might come up with when queried, without much stage setting, for discoveries in environmental studies.

On reflection, we might discount some items on our list for three different reasons. First, the technologies strike me more as inventions than as discoveries. Like created artworks, technologies don't pre-exist their discovery. As such, they don't have the »factivity« that we associate with discovery: it doesn't make sense to ask whether an invented technology (or an artwork) might be false, or not apply to reality in some important manner. By contrast, establishing that a putative discovery is false or non-existent is enough to rule it out as a genuine discovery. So, even though central inventions might be used to characterize a field, they don't speak to the knowledge-generating sense of disciplinarity I'm pursuing here.

By contrast, the causal connections on the list are clearly factive. I can't discover that the reintroduction of wolves caused increased river meandering if this increase is otherwise caused, or if the evidence of increased meandering falls apart. And the factivity condition also applies to the prominent concepts, though a little less obviously. Particular mental episodes take place in the heads of discoverers, but concepts, understood as the multiply instantiable features these episodes of thought concern, seem just as discoverable as new proofs, new colors or even new species of animal. Garrett Hardin (more or less) discovered the tragedy of the commons, for example; and in developing his eponymous formula, Yoichi Kaya discovered it. We might always ask whether it's accurate to apply a discovered concept or model to a given environmental situation, just as there's always a question of whether a given specimen falls under a discovered species. But concepts themselves can be discovered.

A second reason for throwing out some items on our list is that, although the factivity condition applies to them, they fail to satisfy it – the putative discoveries are actually false or non-existent. I can't discover the connection between wolves and rivers if that causal connection doesn't, in fact, obtain. I won't pursue this second reason here, though: I won't defend the accuracy of the concepts on the list, or the truth of the causal connections (though I think they're all true). This is because if these examples fail to be accurate or true, many others could be substituted. My goal here isn't to question whether the

---

10   These formulae model environmental/climate impact as a function of population, rate of consumption, technology, carbon intensity, and so on.

examples count as genuine discoveries, but rather to ask whether they would belong centrally to environmental studies if they did.

The third, and most relevant reason for removing an item from our list is that it belongs to environmental studies only in the broad inclusive sense, not in the more demanding exclusive (»no double counting«) sense we seek here. This concern might apply, for example, to the concept of demographic transition, which is central to the discipline of demography, despite having important applications to environmental issues. We're looking for a discovery that belongs to environmental studies in a way in which it doesn't belong centrally to any of the contributing disciplines. In order to explore how we might understand and legislate this more exclusive condition of »belonging to a field,« it will be useful to pursue just one of our examples in more detail.

10.5   Columbus Caused the Little Ice Age

In 1492, Columbus kicked off a wave of new world exploration and, along with it, the era of globalization. The »Columbian exchange« between the Old and New Worlds sent animals, plants, crops, commodities, people, cultures, weapons, technologies, and much more in both directions. Indigenous Americans got the short end of this exchange, suffering war, conquest, slavery, dislocation, genocide and, notoriously, disease. Waves of lethal Old World diseases (smallpox, scarlet fever, yellow fever, malaria, measles, typhus, and influenza) devastated previously unexposed indigenous populations. Multiple epidemics and deadly conflicts with newly-arrived Europeans, as well as knock-on effects (dislocation, internal strife, famine), combined to reduce the indigenous population by an order of magnitude – literally. Recent estimates are that at the time of Columbus's voyages, the Americas had a population of roughly 60 million people – in the same ballpark as Europe (approximately 80 million) and China (100 million). But after a century, the »Great Dying« had reduced indigenous populations by over 90% – at least twice the death rate inflicted by the Black Death upon 14th century Europe. One reasonable estimate puts indigenous population in 1600 at merely 5 million or so.[11]

This astonishing and catastrophic depopulation had all sorts of side-effects. Many indigenous societies were highly agricultural, so as their populations

---

11   For more precise versions of these estimates, and full explanation see Koch et al. 2019, 14–17. This section draws heavily from this excellent article, as well as from part 6 of Ruddiman 2014. Ruddiman is one of the originators of the causal hypothesis under consideration.

collapsed, fields, farms, terraces, and even fire-tended forests were left fallow. Over the course of the 16th century, as much as 1% of the Americas' land surface area was abandoned and reforested.[12] Since there's typically much more biomass in re-wilded land than there is in farmland, this led to significant carbon uptake across the hemisphere. Carbon sequestration from reforesting tends to be front-loaded, with more of it happening in the first decades after abandonment. And since the use of fire was wide-spread among indigenous populations, their rapid depopulation also quickly *decreased* the amount of anthropogenic carbon put into the atmosphere. So, the depopulation of the Americas plausibly contributed in several ways to a rapid and significant decrease in atmospheric carbon over the course of the 16th century.

Glacial ice core samples confirm a general global drop of approximately 7–10 ppm CO_2 over the century or so following first contact. And this dovetails with the depth of the Little Ice Age, a period of terrestrial cooling throughout the 16th to 19th centuries, during which temperatures dropped below previous means by as much as 2 degrees Celsius. European winters were notoriously harsh during this era, with snow lingering longer, sea ice extending further, glaciers expanding, and rivers and canals freezing over. During the 17th century, regular »frost fairs« were held on London's frozen River Thames.

Figure 10.1   »The Frozen Thames 1677« by Abraham Hondius © Museum of London

---

12   Koch et al., 24.

Although there's some disagreement about exactly when the Little Ice Age began, it's clear that the depth of this climate anomaly occurred during the two centuries or so after Columbus's voyages, with the first of three temperature minima occurring in 1650.

Over the last few decades, a number of scholars have proposed that carbon sequestration in the Americas may have played a significant role in this lowering of global surface temperatures. And accumulating evidence increasingly supports this hypothesis. Exactly how strong a role American carbon uptake played in the Little Ice Age is difficult to estimate, of course, especially since other factors might be in the mix: for example, increased volcanic activity, decreased solar activity, shifting ocean currents, and orbital forcing (small cyclical changes in the tilt of the Earth's axis and orbit around the Sun). Plague-caused depopulation in Eurasia starting a few centuries earlier may also have played a similar role. However, these seem less likely to be competing hypotheses than additional contributing causes of the Little Ice Age. The difficult question is how strong a role should be attributed to each factor.

A sophisticated recent meta-analysis of this complicated causal story concludes that the American carbon uptake contributed the lion's share (roughly a half to two thirds) of the cooling that took place during the Little Ice Age.[13] If we accept the results of this well-regarded study, carbon uptake in the Americas is the most significant cause of the Little Ice Age. And since the cause of this carbon uptake was the Great Dying, which in turn was caused by contact with Europeans, all of this falls squarely on the shoulders of Christopher Columbus, who kicked it all off. So, it's not entirely hyperbolic to say that Christopher Columbus caused the Little Ice Age.

10.6   A Sense of Belonging

Is this a discovery that belongs centrally to environmental studies? Well, to start with, there is still uncertainty about whether the claim that Columbus caused the Little Ice Age is actually true. As We've seen, there are many links in the causal chain, and each link has evidential support that is complex (often statistical). Moreover, the way I've stated the claim negotiates the murky vagaries of causal explanation.[14] Columbus's voyages were not on their own sufficient

---

13   Koch et al., 30.

14   I would definitely break the bounds of causal explanation if I went any further: if I were to claim that the Little Ice Age was caused by a butterfly crawling over a lunch-time orange that (according to what I learned in kindergarten) inspired Columbus's journey; or to

for the wave of New World exploration, of course. Many other factors had to be in place – a general expansionism in European nations, improved maritime capabilities, and so on. Moreover, the voyages were not even on their own necessary: if Columbus hadn't set sail when he did, some other European might well have done so soon thereafter, and with similar effects. And of course, what is perhaps the most surprising part of the story – the links between agricultural abandonment and global atmospheric change – stands alongside the other potential partial causes. Still, given our current evidence, the causal claim is plausible enough for our purposes;[15] even if further investigation shows the claim to be false or exaggerated, we could substitute another example. That Columbus caused the Little Ice Age is the type of fact that would clearly count as a discovery.[16]

---

      claim that Columbus caused the Thirty Year War (which plausibly resulted in part from economic-political instability brought on by the Little Ice Age).

15   In fact, the claim that Columbus caused the Little Ice Age is in certain ways more accurate than the very paradigm of explorational discovery – Columbus's discovery of the New World. The latter is challenged not only by earlier Norse and possibly Basque mariners, but by the pragmatically concealed absurdity of saying that Columbus discovered a hemisphere already inhabited by 60 million Native Americans who weren't always pleased to discover Columbus discovering them. Moreover, there's an intentional mismatch: Columbus didn't take himself to be discovering a new hemisphere, but a new route to India.

16   Of course, what makes for a discovery is itself under dispute – in this very volume, in fact. But our causal claim would seem to count as a discovery on any plausible account. It certainly satisfies Michel's useful and plausible criteria: it's a claim that (i) counts as knowledge; (ii) was found to be true (collectively and over some time); and (iii) has been accepted by the relevant individuals and institutions of academia. Michel proposed these criteria at the Making Discoveries conference. But see also Michel 2019.

     Participants at this conference wondered whether my example is even a candidate for discovery because it's a *singular* causal statement, and, connected with this, it doesn't yield confirmable predictions. I think the first worry places too strong a condition on discovery. For one thing, it threatens to rule out discoveries within the field of history, which traffics significantly in singular causal claims. And even in the natural sciences, we regard certain singular causal claims as significant discoveries – e.g., the Big Bang, and the Alvarezes's claim that 68 million years ago the Chicxulub asteroid wiped out the (non-avian) dinosaurs along with three quarters of plant and animal life generally.

     As for the second worry, we might question whether discoveries need to bring with them predictive power – discovered lands and species don't obviously do this. But even if it's true that singular historic causal claims don't on their own yield predictions, certain historic discoveries are notable precisely because they rest on conceptual insights which themselves have predictive power. (Otherwise, those who don't know history wouldn't be doomed to repeat it.) Thus, the Alvarezes taught us not only about the demise of certain dinosaurs long ago, but also about the effect asteroids can have on terrestrial life in the future. And so too, the claim that Columbus caused the Little Ice Age opens the door to

The claim is also a nice example of collective discovery.[17] In fact, the group epistemology involved can help us see why the discovery might belong uniquely to environmental studies. Notice the astonishing diversity of academic fields involved in supporting the claim: astronomy, paleo-climatology, environmental history, epidemiology, land-use archeology, demography, physical geography and even statistical epistemology. Support for the causal claim comes not just from people working in a variety of well-established disciplines, but from a diverse variety of scholarly methods – everything from ice-core analysis to archival research to statistical modeling.

So, in what sense might this discovery belong centrally to environmental studies? Was it made deploying the methods and principles of environmental studies? No: We've just observed the diverse range of methods and disciplinary contributions that were involved, all of which confirms my earlier claim that environmental studies doesn't have a distinctive methodology.

Was the discovery made exclusively or even primarily by scholars institutionally affiliated with an environmental studies program? No: even if a few researchers did have such affiliations,[18] support for the discovery comes from scholars with all sorts of departmental affiliations. And in any case, the matter shouldn't turn on this type of institutional contingency. I want to allow that the discovery might still count as belonging to environmental studies even if no researcher had an official environmental studies affiliation.

Can we claim, even so, that the discovery wouldn't have been made if it weren't for the academic institutions of environmental studies? No: this also seems doubtful, though the existence of the institutions might have helped. And again, I want to leave open the possibility that a discovery could count as centrally within environmental studies even if it predated the existence of so-labeled institutions. (I'll claim later that the discovery of eco-holism has this status.) For similar reasons no contemporary philosopher would object to classifying Newton's work as squarely part of physics, even if Newton considered himself a »natural philosopher«.

With three strikes against us, then, why – and in what sense – should we nevertheless consider this discovery as belonging centrally to environmental studies? I think there are two connected reasons.

---

      finding and predicting other human effects on global climate. Hence, Ruddiman's »early Anthropocene« hypothesis.

17   See Michel 2019, 430 for insightful discussion of this aspect of many discoveries.

18   Notably Ruddiman, a professor in University of Virginia's Department of Environmental Sciences.

First, the discovery has a synthetic quality that makes it epistemically »over-and-above« the type of facts explored by the contributing disciplines. Even if a philosopher convinces us that our causal claim (like all macroscopic causal facts) is *metaphysically* reducible to patterns of atoms in the void, the claim nevertheless strikes me as *explanatorily* irreducible to any one discipline. By this I mean that the causal claim (and its evidential support) couldn't be comprehensively explained or fully understood by restricting oneself only to the concepts, categories, principles or vocabulary distinctive of just one or another of the contributing disciplines. For example, the causal claim centrally involves the methods and principles of climatology. But a full understanding goes much further. Trying to tell the story entirely within climatology would be like trying to use microphysics to explain why a square peg can't go into a round hole (to invoke Putnam's (1973) example). To which contributing discipline would this discovery squarely belong? There is none. Perhaps history comes closest, but if so, it falls within a species of history – environmental history – that itself draws essentially from other disciplines. (In fact, properly considered, I think environmental history should be seen as a central sub-field of environmental studies.) So, since the discovery doesn't plausibly belong centrally to any contributing discipline, I claim environmental studies as its natural home.

Second, and more positively, the discovery speaks to the central concerns of environmental studies. Moreover, it was driven by these concerns. What I mean by this is that if it weren't for a keen interest in the environment around us, and a deep concern for the ways we affect it, no one would have particular interest in the exact cause of what is, in the geological scope of things, a relatively minor and historically specific variation in mean temperature. Compared to the much more extreme variations in temperature and glaciation that mark full-on ice ages the Little Ice Age doesn't even come close. So why other than our own involvement would we make such a fuss over one of its partial causes?

It's important to emphasize here the normative character of our academic concerns. The claim that Columbus caused the Little Ice Age is most interesting not because it's intellectually surprising (although it certainly is that), but because of the implications it has for our actions. The claim makes a difference to when the »Anthropocene« begins. It suggests that the era of humanity's global influence on the Earth's planetary systems began several centuries before the nuclear era or even the Industrial Revolution. With our discovery established, we might date it to 1492 – the beginning of the Columbian Exchange and advent of true globalization. In fact, the discovery opens the door to starting the epoch 8,000 years earlier, with the dawn of agriculture.[19]

---

19   See Ruddiman 2014.

And when we begin the Anthropocene – indeed, why we even choose to use this label to delimit a period of geologic time – is itself important only because it emphasizes how globally powerful our actions have been, and the level of caution we should be taking moving forward. It's mildly interesting to sort through the causes of a few abnormally cold centuries; it's very interesting and important to learn that we, humans, even without our current population and carbon-intensive technologies, played a central role.

If I'm right, the causal story of the Little Ice Age is a discovery that belongs centrally within environmental studies. So, even if the field lacks unique methods or principles, it's capable of generating genuine new knowledge. And in this important sense, then, environmental studies is like a discipline. My argument contains both exclusive and inclusive strands. I claimed that, because of its synthetic character, this discovery doesn't belong squarely to any of its contributing disciplines; so, it belongs most naturally to the field that comprehends these disciplinary contributions. And I suggested, more inclusively, that the discovery naturally belongs to environmental studies because it results from the field's methodologically diverse and synthetic character.[20] Moreover, it answers to, and was driven by the field's central concerns. But what, more exactly, are these central concerns?

10.7   Ecoholism as Disciplinary Lens

In the last section, I characterized environmental studies as driven by a keen interest in the environment around us, and a deep concern for the ways we interact with it. This broad norm goes beyond the academic field of environmental studies, of course. It's at the heart of environmentalism generally, and felt, at least to some degree, by much of humanity. In this section, I want to unpack this norm: I'll propose that the central concerns of environmental studies can properly be labelled »ecoholism«, and that they constitute a »disciplinary lens« that can usefully unify the field.

Environmental studies is driven, I suggest, by an implicit or tacit concern for the »environment-as-such«. It's grounded in a commitment to the intelligibility of talking »holistically« about features of larger environmental structures,

---

20   Anecdotally, and as a counter to the worry about curricular breadth, graduates from our environmental studies program often report that it is precisely their engagement with the field's broad and synthetic character that serves them well as they enter a variety of environment-related fields – green energy, sustainable food, environmental advocacy and law.

that is, ecosystems: mountain ranges, water-sheds, marshes, forests, deserts, kelp-beds, lakes, or, to go larger and more controversial, Leopold's »the Land« and the Gaia (»mother earth«) of Lovelock and Margulis.[21] This »ecoholist« view has been articulated in different forms by western thinkers such as Aldo Leopold and the deep ecologist, Arne Naess,[22] but I think the view can be found in some way or another in many thought traditions. This is because the view is natural and intuitive; and, as I'll now argue, it needn't entail extravagant theoretical commitments in the way it might seem.

Ecoholist concerns have both a descriptive and a prescriptive aspect. Descriptively, the ecoholist holds simply that it's intelligible and useful to talk about the functions and interactions of individuals within the ecosystems in which they're embedded. This implicit descriptive commitment is central to the field of ecology, certainly, but also, I think, to most other inquiries that fall centrally within environmental studies. Minimally, the commitment is merely that eco-systemic talk gets usefully at the joints along which nature is carved, and that it can do so in a way that is beholden to truth. But this is entirely compatible with a metaphysically reductive picture of nature. On final analysis, ecosystems might be nothing more than complicated arrangements of organisms and the many other macroscopic and microscopic entities that inhabit a portion of space-time; and these in turn might be nothing more than swarms of atoms in the void. If so, then it might well be that our eco-systemic talk traffics ultimately in mere statistical patterns among the things that inhabit different regions of space-time (regions which themselves are somewhat arbitrarily and vaguely delimited). But all of this is compatible with the view that eco-systemic talk is, nevertheless, intelligible, truth-apt and useful – explanatorily, predictively, and normatively.

Of course, we might find a reductive picture of this type implausible, and so hold that our talk is grounded in larger environmental structures that are metaphysically (not just linguistically or explanatorily) »over and above« the individual entities that constitute them. On this stronger view, ecosystems have metaphysically individual-transcendent needs, interests, well-being, and so on. But ecoholism, at least as I'm using the label, doesn't require this stronger metaphysical commitment.[23]

Ecoholism also has a second, prescriptive dimension: the central concerns of environmental studies aren't just intellectual, but also value-laden.

---

21   See Leopold 1970, Lovelock 1979, and Margulis 2008.

22   See Naess 1973 and 2010, in addition to Leopold 1970.

23   In fact, Aldo Leopold himself might have held only to the weaker form of ecoholism. See Varner 2002 for this interpretation.

Sometimes our concern is aesthetic. We find certain ecosystems interesting because we appreciate the causal processes that underlie their delicate equilibria. We marvel at their integration – the often complex interactions and symbioses in virtue of which individuals constitute an ecosystem, and ecosystems define in turn the individuals embedded within them). And sometimes we simply appreciate the sheer beauty of the intricate ecosystems we discover around us.[24] But a virologist might also appreciate – even admire – the intricate workings of the coronavirus, and still think it should be eliminated. So, when it comes to certain ecosystems, our value-laden concerns shade from the aesthetic into the ethical.

We often think, talk and act as if ecosystems can be objects of moral considerability: the existence of certain ecosystems should be promoted, protected, and preserved; and we should honor their needs and interests. It's a fraught issue, of course, exactly which ecosystems should be protected, and how, especially when doing so conflicts with the competing interests of humans, non-humans and other ecosystems. Nevertheless, I think even the most hardened scientist will agree that what motivates an investigation in environmental studies is not merely describing and appreciating the workings of nature (including our role within it), but a moral concern that things should go one way rather than another. We often think there's a better and worse when it comes to our environment, and we feel some moral pull to act on this concern.

This normative dimension might seem the more controversial aspect of ecoholism. But again, I think it's useful to distinguish a stronger, more straightforward form of the view from a weaker, more conventionalist version. The strong ecoholist holds that we value ecosystems not merely for the instrumental value that the environment has for humans – because of ecosystem services, for example, or the aesthetic pleasure we derive from natural settings. The strong ecoholist holds that environmental wholes have intrinsic value in their own right, with individual-transcendent interests that should be in the normative mix, alongside those of individual humans (as well as, depending on your view, those of sentient non-human animals, and even »merely living« organisms). The strong ecoholist regards environmental structures much in the top-down way a doctor might regard her patient's body – the well-being of the body as a whole trumps the interests of particular bodily parts. In a trade-off between the health or even future existence of a spleen or an arm and the health of the patient's body, the doctor always chooses the body.

---

24   See the articles in Carlson & Berleant 2013 for more about our aesthetic appreciation of
      nature.

But a weaker form of ecoholism is more deflationary: talk of the needs and interests of an environmental whole is just a useful *facon de parler* – shorthand talk about the summed interests of the individuals that constitute an ecosystem. The weak ecoholist values an ecosystem in the way in which a social club might be valued by its members. It's bad if the club goes into debt – but only if and because its members will collectively and individually suffer. If the members willingly want to disband, the club's termination is no moral harm. It's in this way that the weak ecoholist regards the needs and interests of an ecosystem.

My central proposal here is that ecoholism in some form serves, or should serve, as a »disciplinary lens« for environmental studies.[25] By this I mean that »ecoholism« usefully labels a constellation of descriptive and prescriptive[26] concerns – a way of regarding the world – that motivates and binds together the scholarly activities that take place within environmental studies. My proposal gets support by reflecting on what takes place within institutions of environmental studies, as well as the implicit intentions of the people who establish and sustain these institutions. But my claim isn't just a description of an intellectual focus implicit within some frozen or static field of study; it's meant to be a proposal about how the field of environmental studies should see and shape itself as it matures. My goal in discussing different aspects and varieties of ecoholism has been to demonstrate the types of more precise commitments that might lie behind what can, nevertheless, remain an implicit and somewhat indeterminate set of concerns.

I want to sharpen my proposal further by addressing several worries one might have about whether ecoholism can serve as a disciplinary lens for environmental studies. First, many scholars – perhaps most scholars – working in environmental studies haven't heard of »ecoholism«. More importantly, they wouldn't readily agree to the view once it's explained, much less that it drives and unifies their scholarly activities. I have no doubt this is true. But this is compatible with my claim that ecoholism is *implicit* in some way in our motivations and institutional conceptions, and revealed once we reflect upon them. This is in much the same way that most citizens of a liberal democracy are implicitly committed to the importance of equal rights and justice, even

---

25   So as to avoid a terminological sleight of hand, I apply the label »disciplinary lens« in a
      manner that doesn't, on its own, imply disciplinarity. I return to the question of whether
      environmental studies counts as a discipline in the final section.
26   Chapman 2007 suggests that, at least curricularly, the normative dimension of envi-
      ronmental studies should be central and primary. I think this overplays the normative.
      Too bad though, since Chapman goes onto argue that environmental studies should be
      regarded as a form of applied philosophy!

if they've never articulated to themselves a clear conception of these rights, much less a procedure for weighing them against one another and the common good.

One role of the theorist is to articulate, sharpen and sometimes tweak our implicit views. And so, the point of unpacking ecoholism above is to show the variety of explicit and sharpened views that might lie behind our shared, implicit and indeterminate commitment to ecoholism. Sometimes we might think like a strong ecoholist (I know I do), but we might assent only to weaker forms of the view when the issues are laid out before us. In fact, though, we needn't ever make the implicit explicit – we needn't worry about the philosopher's distinctions and niceties – in order for it to be true, nevertheless, that a broad ecoholistic vision guides our activities.

Second, in proposing ecoholism as a disciplinary lens, I don't claim that *every* scholarly activity – every contribution and contributor to new knowledge that falls squarely under environmental studies – is shaped by this lens. Research is a collective and sometimes serendipitous enterprise; and scholars have a fierce independence of mind that willfully resists categorization, especially institutional categorization. My claim is only that ecoholism captures what is *central* to environmental studies.

I don't mean my proposal to be exclusionary or inflexible. I mean it to allow that the disciplinary lens of environmental studies might be modified or supplemented over time. Indeed, environmental justice (»EJ«) has emerged as an increasingly important area of study within environmental studies. And one might wonder how this more sociological and politically-based focus can be squared with the less anthropocentric lens of ecoholism. In fact, I would argue that what distinguishes environmental justice from inquiries into social or political justice more generally is the way in which ecoholistic concerns lurk in the background of environmental justice. Still, there's no reason that we couldn't come to count environmental justice as a second, related disciplinary lens in what would then become a binocular field.

Third, my proposal plausibly allows a certain amount of seemingly anachronistic appropriation. It would be a stretch to call Aldo Leopold a scholar of environmental studies, since Leopold predates the labels and institutions of the field. (Leopold firmly and happily considered himself a professor of game management.) Still, much of his scholarship might be claimed (even if nonexclusively) as belonging to environmental studies because it was motivated and answers to the field's disciplinary lens. Leopold is easy, of course, since one of his discoveries – the Land Ethic – is a paradigmatic version of my proposed disciplinary lens itself. But Rachel Carson's seminal *Silent Spring* from 1962, or even (some of) the work or the 19th-century American writers Henry David

Thoreau and George Perkins Marsh might, in the same way, be said to belong in the field.

Fourth, my proposal invokes the general notion of a »disciplinary lens«. My idea here is that a field might be centrally bound and unified by a theoretical lens – a general outlook that has a descriptive and/or prescriptive focus – even if it lacks unifying principles or methodologies. And my claim is that a disciplinary lens gives a field more than just a well-defined topic; it articulates the *way* that this topic is to be investigated, and the reasons for doing so.

Finally, one might worry that the lens I propose – ecoholism – is simply implausible. And indeed, ecoholism faces some significant empirical and conceptual challenges. Even if one can draw empirically useful spatial and temporal boundaries for ecosystems, and talk sensibly about their needs and interests, one needs reasons to privilege certain types of ecosystems in acts of preservation, sustainability and restoration. Appealing to what's »natural« goes only so far, since, even when untouched by humans, nature is characterized as much by disturbance, flux and open-endedness as by regular succession towards balanced and stable equilibria. More normatively, ecoholism confronts the charge of »eco-fascism« – the objection that caring for holistic entities comes at the peril of the rights and well-being of individuals (humans and, perhaps, sentient critters). Perhaps concern for eco-holes needn't supplant, but can rather pluralistically supplement our concern for individuals, but if so, then one needs a way of weighing the seemingly incommensurable needs and interests of an expanded bestiary of moral objects. Finally, it's unclear what type of positive argument might be given for ecoholism, other than that we sometimes think and act as if it's true.

As I say, these are significant challenges. But although I won't try to do so here, I do think something can be said to meet some of these worries, especially if ecoholism can be sharpened and qualified well-beyond the brief sketch I've provided here.[27] Beyond this, I want to suggest – controversially, I know – that a disciplinary lens might usefully ground a field and lead to new knowledge even if it can't, at the moment, meet all such challenges.[28] So too, ecoholism might usefully bind and rally environmental studies to important discoveries, as well as important activism, even if certain parts of the view remain unresolved. A

---

27 See Callicott 2007 and 2013 for an attempt to address some of these challenges as they apply to Leopold's Land Ethic. See Sober 1986 for additional worries, as well as Pickett & Ostfeld 1995.

28 Although I won't do so here, I would argue, even more controversially, that a disciplinary lens might fruitfully guide a field, even if it is off-target in certain ways. Consider well-functioning fields like psychoanalysis, holistic medicine and theology, which might well have ideological commitments that are simply false.

fruitful analogy here might be with the field of medicine. Medical researchers would all agree that the intellectual project of understanding the workings of the human body is entirely driven by the goal of improving and promoting human health, even if they disagree about which medical interventions are best, or about what exactly constitutes human health in the first place.

It's for all these reasons, then, and with all these qualifications, that I propose ecoholism as the disciplinary lens for environmental studies – as capturing the approach to our environment and its health that drives the field.

10.8   Whither Disciplinarity?

I started this paper by asking whether environmental studies is an academic discipline. I discussed the reasons why this question might matter, and also some ways in which the field of environmental studies differs from traditional disciplines. I left this question unresolved as I pursued the related question of whether environmental studies might generate new knowledge in the form of discoveries. About this second question, I suggested that the capacity to generate discoveries would make environmental studies importantly discipline-like, and I argued that there is at least one discovery – Columbus caused the Little Ice Age – that belongs centrally to environmental studies and no other field. Reflecting on this discovery and others like it led me to articulate and propose ecoholism as the field's unifying disciplinary lens – as the reason why this and other bits of new knowledge should count as belonging to environmental studies.

But what of my starting question: have I established through all of this that environmental studies is a proper academic discipline? The answer, I'm afraid, can only be as determinate as our concept of a »discipline«. We've seen that environmental studies lacks central laws, principles or a shared methodology. In this way, it's unlike traditional paradigmatic disciplines. On the other hand, if I'm right, environmental studies is, on its own, importantly capable of generating new knowledge in the form of discoveries. Moreover, in ecoholism the field has an implicit unifying lens. And these, along with its growing institutional status, are surely central aspects of disciplinarity.

In the end, it doesn't matter whether or not environmental studies is fully and properly called a »discipline«. What's more interesting is to have its contours revealed by making explicit the similarities and dissimilarities it bears to traditional disciplines. This is what I've tried to do here. And I regard this as a useful step toward, as it were, institutional self-realization. With its contours more fully in view, environmental studies can more mindfully embrace its distinctive nature: environmental studies is at once an importantly interdisciplinary field, but also one with a unified and knowledge-generating sense of purpose.[29]

References

Callicott, J.B. 2007. »The Land Ethic.« In *A Companion to Environmental Philosophy*, edited by D. Jamieson, 204–217. Malden, Mass.: Blackwell.

Callicott, J.B. 2013. *Thinking Like a Planet: The Land Ethic and the Earth Ethic*. New York: Oxford University Press.

Carlson, A. & Berleant, A. 2004. *The Aesthetics of Natural Environments*. Peterborough, Ont.: Broadview Press.

Carson, R. 1962. *Silent Spring*. Cambridge, Mass.: Houghton Mifflin.

Chapman, R.L. 2007. »How to Think about Environmental Studies.« *Journal of Philosophy of Education* 41 (1): 59–74.

Foucault, M. 1995. *Discipline and Punish: The Birth of the Prison*. 2nd ed. New York: Vintage Books.

Koch, A., Brierley, C., Maslin, M.M., & Lewis, S.L. 2019. »Earth System Impacts of the European Arrival and Great Dying in the Americas after 1492.« *Quaternary Science Reviews* 207: 13–36.

Leopold, A. 1970. *A Sand County Almanac: With Essays on Conservation from Round River*. New York: Oxford University Press.

Lovelock, J. 1979. *Gaia: A New Look at Life on Earth*. Oxford: Oxford University Press.

Margulis, L. 2008. *Symbiotic Planet: A New Look at Evolution*. New York: Basic Books.

Michel, J.G. 2019. »How Are Species Discovered? Declarative Speech Acts in Biology.« *Grazer Philosophische Studien* 96 (3): 419–441.

Naess, A. 1973. »The Shallow and Deep, Long-Range Ecology Movement: A Summary.« *Inquiry* 16 (1): 95–100.

Naess, A. 2010. *The Ecology of Wisdom: Writings by Arne Naess*. Berkeley: Counterpoint Press.

---

29   This paper is based on a talk delivered at the wonderfully interdisciplinary Making Scientific Discoveries conference, held in Münster, Germany in December 2019. Many thanks to the conference participants for useful comments and suggestions, and especially to the organizer, Jan Michel, who inspired me to pursue this project in the first place. Thanks also to colleagues at Amherst College, especially Ted Melillo.

Pickett, S.T.A. & Ostfeld, R.S. 1995. »The Shifting Paradigm in Ecology.« In *A New Century for Natural Resources Management*, edited by R.L. Knight, & S.F. Bates, 261–278. Washington, DC: Island Press.

Putnam, H. 1973. »Reductionism and the Nature of Psychology.« *Cognition* 2 (1): 131–146.

Ruddiman, W.F. 2014. *Earth Transformed*. New York: W.H. Freeman.

Snow, C.P. 1959. *The Two Cultures and the Scientific Revolution*. Cambridge: Cambridge University Press.

Sober, E. 1986. »Philosophical Problems for Environmentalism.« In *The Preservation of Species: The Value of Biological Diversity*, edited by Bryan Norton, 173–194. Princeton: Princeton University Press.

Soulé, M.E. & Press, D. 1998. »What Is Environmental Studies?« *BioScience* 48 (5): 397–405.

Varner, G.E. 2002. *In Nature's Interests? Interests, Animal Rights, and Environmental Ethics*. New York: Oxford University Press.

New Findings in Old Religions? On the Possibility of Scientific Discoveries in Theology

Christian Tapp

11.1 The Question of Scientific Discoveries in Theology

11.1.1 *Thesis of this Paper*

The guiding question of this paper is whether there are scientific discoveries in theology[1] and, if so, what their nature is. Can there be new findings in old religions?

One might think there cannot. One might think that religions, as repositories of eternal truth, can do no more than repeat the same truth claims over and over again. One might think that religions are morally unteachable, so that they adhere to old-fashioned moral principles, ignore secular ethical insights, and presume that their old answers fit even the newest ethical problems. Finally, one might think that the identity of a religion so depends on certain historical documents or events that it requires that religion painstakingly to perpetuate the contents of a certain body of holy scriptures, revelations, wisdom sayings, or creeds.

One might think all of this. And these thoughts certainly contain some grains of truth. But taken as the expression of a position that implies the impossibility of discoveries in theology, they are false. On the contrary, I will show in this paper that for some religions a self-understanding is accessible that allows for scientific discoveries in associated theologies.

To be clear: this paper is not about discoveries in religion in general but in theology.[2] By the theology of a certain denomination I roughly understand the academic field in which its religious beliefs are studied in various respects,

---

* I am indebted to Jan Michel, Christian Weidemann and participants in the »Making Scientific Discoveries« conference held in Münster, December 3–5, 2019, for many valuable suggestions.

1 This paper is largely written on the basis of reflections concerning Catholic theology. This is sometimes reflected in the terminology used (e.g., »moral theology« or »fundamental theology«). Most considerations will probably hold *mutatis mutandis* for Evangelical theology, some of them even for non-Christian theologies.

2 This book was originally to include a separate paper on new developments in the non-theological aspects of religions. Unfortunately, that paper was ultimately not submitted for publication. – Concerning the question of new developments in dogma, see also my 2015.

© BRILL MENTIS, 2022 | DOI:10.30965/9783957437044_012

including their critical interpretation, their discussion »in the light of reason,«
and the examination of the perspective »from within,« i.e., under the presupposition that they are true.[3]

The word »science« is used throughout this article in a broad sense that includes the social sciences and humanities. In English, this is an old and somewhat outdated meaning of the word »science,« which is now almost exclusively used to refer to the natural sciences. I will nevertheless stick to the old, broader meaning. This may be slightly uncomfortable for contemporary English readers, but I know of no better alternative. No other general term in English carries all the implications, presuppositions, norms, and expectations shared by the rigorous and methodical academic disciplines that in German are called »*Wissenschaften.*«

11.1.2   A »*Philosophy of Theology*«?

The nature of scientific disciplines, the criteria demarcating science from pseudo-science, and the existence and nature of scientific discoveries are topics that belong to philosophy of science broadly conceived.[4] Accordingly, philosophy of science is the first place to look for answers to the question of the existence and nature of scientific discoveries in theology. Is there such a thing as a »philosophy of theology« in the sense in which there is a philosophy of mathematics, a philosophy of biology, and a philosophy of the social sciences?

In terms of recognized professional specializations, there is no philosophy of theology, at least not at a level comparable to philosophy of science generally. However, there is ongoing discussion and a state of the art. A milestone was the famous and groundbreaking 1931 paper »Wie ist eine evangelische Theologie als Wissenschaft möglich?« (»How is evangelical theology possible as a science?«) by Heinrich Scholz.[5] In that paper, Scholz developed criteria that, if met by theology, would establish theology's status as a truly scientific field. And Scholz judged that the criteria could indeed be met. In the 90 years since then, philosophers have discussed many different criteria for whether an activity is scientific and have raised several critical problems for the scientific

---

3   This is a very rough characterization of academic theology. It does not reflect all or even the most relevant aspects of the field, as will become clear when some characteristics of theology's subdisciplines are discussed below. Still, it may be helpful as a first approximation for readers who are unfamiliar with academic theology.

4   By »philosophy of science broadly conceived« I mean the whole area of historically informed philosophical reflections about the academic fields (in the sense of the German »*Wissenschaftsphilosophie*«). It includes what is sometimes called »general philosophy of science« as well as the philosophies of particular sciences, of the humanities, and so forth.

5   Scholz 1931.

status of theology.[6] Most of these problems have not yet been answered satisfactorily. If answers have been proposed at all,[7] they have been partial or not fully convincing.

Despite three important monographs in the mid-1970s – by Gatzemeier, Peukert, and, most notably, Pannenberg[8] – the literature has been largely silent on this question, and the philosophers' critical questions have gone more or less unanswered. Since the 2010s, some efforts have been made to push the discussion forward.[9] However, progress on the question whether theology is a scientific discipline is often immobilized, caught between people who feel embarrassed by the question because they are completely convinced that it is and people who feel embarrassed by the question because they are completely convinced that it is not. Certainly, in comparison with theology itself, the philosophy of theology is still in its infancy. But, as we have seen, some important steps have been taken.

11.1.3 *Two General Problems for »Deductive« Solutions*

This sobering diagnosis with respect to a philosophy of theology has implications for the guiding question of this paper (whether there are scientific discoveries in theology and, if so, what their nature might be).

In general, the ideal, »deductive« way to answer such a question would be, first, to establish a concept of the discipline in question, next, to justify regarding the discipline so defined as a scientific one, and, finally, to analyze the nature of the discoveries that are – or can be – made in that discipline. With respect to theology, however, such a procedure faces severe difficulties right from the outset. I will discuss two problems so fundamental that they constitute serious obstacles to answering our question.

The first problem results from the view that belonging to a scientific discipline is what makes a discovery a *scientific* discovery.[10] According to this view, looking for scientific discoveries in a field presupposes that the field in question is indeed a scientific discipline. However, the status of theology as a

---

6   For example, see Morscher 1974; Morscher 2003.

7   For example, see Wendel 2002; Werbick 2010.

8   Gatzemeier 1974–1975; Pannenberg 1973; Peukert 1978.

9   See the results of the Emmy-Noether research group led by Benedikt Göcke in Bochum, especially the trilogy *Die Wissenschaftlichkeit der Theologie* (»The scientific character of theology«), Göcke 2018–2019.

10   As Jan Michel (2020) has argued, it is not necessary to take this view. If it is taken, the problems discussed in this paragraph arise from it. If it is not taken, those problems are replaced by the question of general criteria for demarcating scientific from non-scientific discoveries in a field. Is such a path of inquiry easier to pursue?

scientific discipline is in doubt. Some reasons for this doubt parallel the reasons for doubting the scientific status of cultural studies and the humanities in general (the fields lack empirical foundations, clear distinctions between data and theory, decisive criteria for progress, objectivity, and so forth). I skip over these doubts in the context of this paper because they are not specific to theology. Much more specific are doubts originating from theology's ties with religion: Does a discipline have the necessary freedom of thought if it is required to stay within the boundaries set by a religion? Is the discipline sufficiently independent of non-academic institutions? Can there be progress in research if what is held to be true is largely predetermined? Can a religious person distinguish between truth as scientists seek it and his or her personal religious beliefs about what is true?

A familiar way to allay these doubts is to replace theology with a form of religious studies. Religion is thereby confined to the role of an object of empirical and/or historical studies. However, despite offering many valuable insights into the facts as to how religions are structured, the functions of religions in society, and their historical development, religious studies cannot fully replace theology. For by making such a substitution, we would lose important aspects of what makes rational theology desirable for religious communities as well as for enlightened societies. For both, it is of great value to know exactly what the religions teach, how their teachings shape their believers' world view and moral outlook, and how a productive discourse between secular and religious voices is possible. Secular societies do have a genuine interest in knowing what the »contents« of a religion are: what do the adherents of a certain religion believe and why?[11] How strongly do certain beliefs bind the faithful, and where is room for discussion or compromise? What do religious beliefs contribute to the social life and moral conduct of religious people? Moreover, secular societies have a strong interest in these beliefs being considered in the light of reason and in the context of other academic disciplines with their standards of rationality, yet without alienating religious believers from these rationally developed versions of their faith. And secular societies have an interest in the academic education of ministers, teachers, and other religious officials. For the state and for society, rational religious practice is better than irrational

---

11   To be sure, religious studies can contribute to answering the question what religious sentences religious people assent to. But I take it to be a lesson from holism in philosophy of language that one cannot finally tell what sentences mean independently of a whole belief system, i.e., a network of assumptions concerning the truth of such sentences. Whether religious sentences are true is, however, a question that religious studies brackets. Hence, for an account of what religious believers believe, one needs, in the end, a theological standpoint, even if a hypothetical one.

practice, openness to rational discourse is better than reclusiveness, and the flexibility to make distinctions in religious matters is better than intransigent commitment to unreflectively adopted »dogma.« For these reasons, the proprium of theology – rational reflection about and within a belief system – is of central importance for modern societies. But these considerations are precisely what block the easy way out of our first problem. Transforming theology somehow into religious studies may dissolve some of the doubts about the scientific nature of theology, but it does so at the cost of losing many of theology's benefits for the religious community *and* for secular society.

The second problem is that there is no clear conception of academic theology to rely on. True, theologians use big words to describe their particular ways of doing theology. There are phenomenological and hermeneutical approaches, there is narrative theology and analytic theology, there are post-colonial, post-structuralist, post-modernist, post-metaphysical, and post-you-name-it theological *Ansätze*. While almost all these approaches agree that religious beliefs must be assigned a theological role, there is no clear – or, more to the point, no uniform – conception of how that is to be done. There is no clear methodology as there is in natural science where the inductive method, the interplay between data and theory, the Hempel-Oppenheim schema of scientific explanation, and the like, provide good first approximations of scientific practice – despite their shortcomings. In philosophy of science, conceptual models of scientific disciplines afford at least a preliminary grasp of their nature. But there is no such model that would even broadly unite the most important ways to conceive of theology. To be sure, there are some proposals as to how theology can and should be conceived of if it is to be a truly academic field.[12] But they come closer to constituting a list of criteria for a (possible) scientific theology than to subsuming (actual) theological practice. Hence, the second problem is that there is no clear, even if approximate, unifying conception of theology in general.

11.1.4 Outline of a More »Inductive« Approach

My preliminary conclusion from the above considerations is that the prospects for answering my question »deductively« on the basis of a given conception of theology are poor. This paper simply cannot solve the intricate questions of whether theology is a truly scientific enterprise and whether there is such a thing as a unifying concept of theology. Hence, it cannot use the »deductive« approach outlined above to answer the question whether all theological discoveries are truly *scientific* discoveries.

---

12   See my 2018.

What we can do is to proceed in a more or less *inductive* way by examining several examples of (possible) theological discoveries. We will explore their nature and suggest some arguments for counting them as scientific. My argument for the existence of scientific discoveries in theology and my attempt to clarify their nature proceeds in three steps.

The *first step* requires taking into account that theology is a tremendously interdisciplinary discipline. Academic theology is usually organized as a conglomerate of a dozen or so subdisciplines that can be grouped into four or five fields:
- Biblical theology: exegesis of sacred Scripture (Old and New Testament studies)
- Historical theology: Church history and history of theology (including patrology and archaeology)
- Systematic theology: fundamental theology, dogmatic theology, and moral theology (including ecumenical studies and Christian social ethics)
- Practical theology: pastoral theology, liturgy, canon law, and religious pedagogics.

In a typical theological course of studies, students are also required to take courses in a fifth field, namely philosophy. Thorough training in all five fields constitutes the standard for academic education in theology.

This internal structure of theology has consequences for the existence and nature of theological discoveries. Their nature varies according to the methods used in the various subdisciplines. Most theological subdisciplines have secular counterparts with which they share almost all their methods. For instance, the counterpart of Church history is (general) historical studies, the counterpart of canon law is jurisprudence, and the counterpart of moral theology is ethics. Hence, I argue in this first step that many discoveries in theological subdisciplines deserve to be considered scientific because, essentially, they are made in the same way as discoveries in their counterpart non-theological academic disciplines.

Generally speaking, most theological subdisciplines have more in common with their secular counterparts, such as history or philology, than with the other theological subdisciplines.[13] This raises the question whether theology is more than a conglomerate of subfields of other disciplines. Are there genuinely theological discoveries? I will give a preliminary answer to this question by indicating what may constitute a »theological plus« in the epistemic business of a subdiscipline, i.e., what may distinguish a theological

---

13   In the case of Church history, this thesis is widely held. See Wolf & Seiler 2004, 382, or
      Leppin 2014, 73.

subdiscipline from the corresponding subdiscipline of a non-theological field.

Church history, for example, is first and foremost a field of historical studies. Its methods are almost identical to the usual methods in other fields of historical studies. The only two differences are (a) a specialization in topics concerned with or relevant to the Church or Christian religion in general, and (b) the purposes for which the historical studies are pursued, namely as part of the self-reflection of a particular religious community.[14] Similar considerations hold, to give a second example, for biblical exegesis. In exegesis – at least when it has taken to heart its historical-critical lessons – the biblical scriptures are studied as ancient texts using all the methods of philology and literary studies. The two main differences between philology and literary studies on the one side and biblical exegesis on the other are, again, (a) the specialization or restriction to texts from the Bible (or closely related literature) and (b) the hermeneutical end of the exegetical studies, namely, to elaborate a comprehensive account of biblical theology as an offering to contemporary systematic theological accounts as well as to preaching, catechesis, and the like.

Sections 3 to 5 below will discuss several possible examples of theological discoveries that can quite easily be acknowledged as scientific on the grounds of the methods used to arrive at them.

The *second step* of my argument concerns the central systematic-theological subdisciplines of fundamental theology and dogmatics. These subdisciplines require special attention because they do not have clear secular counterparts. Discoveries in these subdisciplines cannot easily be compared to discoveries in other academic fields. Their nature is what I call »hermeneutical« or »doctrine-related.« Possible discoveries in those fields crucially depend on the underlying conception of theology. According to the conception I have proposed elsewhere,[15] systematic theology should be considered a theory-oriented, hermeneutically proceeding undertaking with religious beliefs playing a role roughly analogous to the role of data in natural science. I do not want to claim that this is the only valid conception of systematic theology. What I do claim is that the more (systematic) theology conforms to this conception, the better its chances of being accepted as a truly scientific enterprise by the scientific community. We will discuss the hermeneutical nature of such discoveries in Section 6.

The *third and final step* of my argument returns to the question of genuinely theological discoveries. Is there something uniting the epistemic undertakings

---

14 The purposes of Church history determine both the heuristic and interpretational steps of its work.

15 See my 2018.

of the theological subdisciplines? Are there any other ways for a discovery to be genuinely theological besides the ones indicated by the »theological plus« in some subdisciplines and the ones called »hermeneutical/doctrine-related« in the central areas of systematic theology? The sheer complexity of the epistemic structure of theological discourse makes this question hard to answer. Nevertheless, I will try to sketch an example of such a genuinely theological discovery in Section 7.

11.1.5   *A Preliminary Classification of Theological Discoveries*

One can distinguish at least four kinds of discoveries in theology (note that the following four kinds of discoveries are related to but not identical with the four sub-disciplinary fields we have just discussed). There are:

1. Historical discoveries
2. (Other) pure and impure empirical discoveries
3. Action-oriented discoveries
4. Hermeneutical or doctrine-related discoveries.

In Sections 3 to 6 below, I will explain what I mean by these kinds of discoveries and present some examples of each of them.

11.2   Conceptual Clarifications and a Caveat

11.2.1   *On Discoveries*

Since this paper deals with discoveries, I should outline what I understand by a discovery. To begin with, »discovery« is an *epistemic* expression. Making a discovery requires an epistemic subject (be it a natural person, a group of persons, or anything else that can count as an epistemic subject) with a certain set of epistemically relevant propositions – typically, the person's knowledge or set of beliefs.

We can discover entities of many different kinds: facts (that this or that is or can be the case), things (a new planet), concepts (such as the biological concept of fish), relations (such as logical relations between certain assumptions of a scientific theory), and maybe even contents or thoughts (that which the sentence »We live in a Matrix« expresses, independently of whether it is true). In what follows, I will chiefly focus on what I take to be the most basic meaning of »discovery,« namely, the discovery of facts. At some points, however, I will discuss the discovery of things, relations, and so forth.[16]

---

16   The question of how the different meanings of »discovery« are related leads deeply into intricate metaphysical questions such as: What does it mean to have a concept? Is there

If, for a proposition p, someone discovers that p, p's truth must be new to the epistemic subject. It must not have been part of her knowledge before now. Furthermore, it must be possible to integrate p to some degree into the subject's body of established knowledge. We would not say somebody discovered that p if p had no logical or linguistic relations to what the person knows. Moreover, making a discovery *requires truth*: I can only discover that p if p is the case. If it were not the case that p, then for trivial reasons I cannot discover that p. In other words: »discovering that« is a *factive* phrase – it can be applied truthfully to a proposition only if the proposition itself is true. Therefore, it is *logically* impossible to discover that 164 is a prime number.[17]

A discoverer may act either as an individual or as a representative of a group (such as the scientific community). Hence, one can distinguish between *two senses* of »discovery,« a *subjective* and an *objective* one. In the subjective sense, individual epistemic subjects act for themselves. In this sense, every student makes many discoveries during her course of study. She discovers everything she learns: facts, concepts, theories, relations between them, and the like. This does not mean that the student ever makes a scientific discovery in the stricter, objective sense of finding out things that are new not only to the discoverer but also to the scientific community, be this the set of all scientists in the past and the present, the whole scientific community at a time, or the community engaged in a particular scientific discipline. Distinguishing subjective from objective discoveries is not a binary matter. There are intermediate cases, such as when something is new to a particular working group or to the subcommunity of people who speak a certain language.

In what follows, I will focus chiefly on the question of objective discoveries. Granted, it might be rewarding from a psychological or pedagogical standpoint to study the individual effects of subjective discoveries, such as the warm feeling of mastery that accompanies such a discovery and the motivational implications such experiences have for one's general attitude towards learning. My focus, however, will be on scientific discoveries that are new to the community of scientists in a given field.

Discoveries can be made on more than one level. One might distinguish between *object- and meta-level* discoveries. On an object level, one discovers facts – for example, that a certain meter displays a certain value or that there

---

more to discovering a thing of a kind f than discovering that there is a thing of kind f? And so forth.

17   Here, I use »logically« in the broad sense others call »analytically.«

is still a Sumatra rhino in Malaysia.[18] On a meta-level, one discovers relations between facts and other facts or between facts and theories. Both kinds of discoveries are epistemologically significant, but the two differ in structure. For example, many facts can be discovered simply by observation, although sometimes only with the help of instruments. And many facts can be articulated within a very rudimentary theoretical framework. The theory is needed only to fix the meanings of the relevant concepts and guide the relevant observations. In contrast, meta-level discoveries tend to require a deeper knowledge of relevant theories. Often they do not require direct access to objects of experience: a typical discovery by a theoretical researcher brings to light, not new object-level facts, but, for example, relations between a fact and its theoretical background.[19] For the discovery of such relations, it is not important whether a fact actually obtains or is only a possibility, whether it is a *Tatsache* or a mere *Sachverhalt* (in Wittgensteinian terms). In a nutshell, facts can be discovered without access to full-blown scientific theories, but relations cannot. Relations can be discovered without direct access to the world, but (empirical) facts cannot.[20] To keep things simple, I will use the same language for both kinds of discoveries, so that *p* in »discovering that *p*« may be an object- or a meta-level proposition.

11.2.2 *Caveat: Idealization*

The reality of academic activities is complex. Most fields are not monopolized by a single accepted method. In many cases, multiple methods are combined. Sometimes which methods are used is not even clear, as researchers may apply methods intuitively on the basis of training rather than of explicit deliberation. Boundaries between subdisciplines are permeable. In theology, not all products of academic theologians qualify as scientific in my view. For example, religious narratives may be constructed with a high level of intelligence and sophistication. But it is more than doubtful whether they should count as scientific, although they may have greater persuasive power, greater influence on society, and/or greater general impact than many genuinely scientific works. Where exactly lies the boundary between an elegantly written, genuinely

---

18   Sumatra rhinos are now considered extinct in Malaysia, since the last known specimen died on November 23, 2019. See https://www.bbc.com/news/world-asia-50531208.

19   For reasons of simplicity, these relations (and not relations on the object level) are referred to as »relations« in the following.

20   Mathematical and logical facts are a special case. I would tend to call them »object-level facts,« but their discovery does not require access to the empirical world. It may be that they are structural in nature and therefore are in reality meta-level facts that merely seem to be object-level facts.

scientific paper and a religious essay that incorporates scientific insights without being scientific in any stricter sense of that word?

These ambiguities notwithstanding, I hope that the idealizations underlying my paper help bring to light some significant characteristics of theology and its subdisciplines. One should, however, keep in mind that reality is much more complex than the idealized image I am working with.

11.3   Historical Discoveries

Among the theological disciplines, there are some in which ordinary empirical discoveries are made. For the most part, such empirical discoveries are historical discoveries.

For example, a Church historian searches the secret archives of the Vatican and finds evidence that leading cardinals were involved in a certain scandalous event. In the library of some ancient monastery, a liturgy professor tracks down a 7th-century missal with glosses that clearly indicate a liturgical role for female deacons. An Old Testament scholar runs through the Qumran fragments and discovers a previously unrecognized fragment of the book of Esther.

These are examples of (possible) historical discoveries in theology. All of them concern historical facts, i.e., the contents of specific historical sentences, sentences about single events or particular persons in the past. History, to be sure, goes vastly beyond collecting facts about the past. Roughly speaking, it explains these facts by integrating them into a coherent narrative. Such narratives are presented in texts that connect historical facts with general insights, offer hypotheses about causal links between events and repeating patterns of such links, and explain the facts in the light of these universal statements. Although historical explanations differ in several respects from explanations in the natural sciences, the resemblance is so strong that one might draw an analogy and speak of »theory« and »data« in history. Historical narratives have both relatively universal and relatively singular aspects. One might call the more explanatory, universal aspects »theories« and the more singular aspects, which are subject to explanation, »facts« or even »data.« Of course, there are no »pure facts.« Facts are available for consideration only under a description. Descriptions make use of general terms under which particulars are subsumed. This subsumption introduces a degree of interpretation into even a simple description. The same is true in natural science. As the famous philosopher W.V.O. Quine has shown, there is no clear boundary between observation sentences and theoretical sentences. Observation is always theory-laden.

Nevertheless, the distinction between data and theory is a helpful idealization, and a similar distinction can be applied to the case of historical studies.

As to discoveries, I hold that, in theology, there are discoveries not only of historical facts but also of historical theories and of relations between facts and theories. A discovery including theoretical aspects might occur when a Church historian writes an original, comprehensive history of certain events, such as the Second Vatican Council. Another discovery of that sort would be involved when a Church historian, for the first time, convincingly explains the irritating politics of a certain former pope. One might reasonably say that this historian has *discovered* that explanation. If one tends, instead, to conceive of such explanations in terms of inventions, one may at least say that she has *discovered that there is* an explanation of the kind she has invented.

Generally speaking, the discovered relations between theories and facts are established by placing singular events in the context of a general historical narrative. In this way a Church historian presents explanations for singular events, i.e., she establishes certain relations between these events and the more general elements of her historical outlook, such as principles, generalizations, hypotheses, or accounts. Relations within one or between two or more historical theories are discovered when, for example, a certain pattern of historical explanation is tested in the context of several general accounts. Our Church historian might discover that the explanatory patterns that worked well in her history of 13th-century France do not work equally well in her history of 14th-century England.

So far, I have strongly relied on the parallel between secular history and Church history. Most contemporary Church historians see themselves mainly as historians. The scientific methods they use belong to the general historians' toolbox. In most of their everyday work, Church historians apply general historical methods to their field of specialization, the history of the Church.[21]

---

21   Or of Christianity, or of theology... I favor the phrase »Church history« only because it is the established technical term for this theological subdiscipline. – For support of this position, see Wolf and Seiler 2004. While there is extensive consensus today that the methods of Church history are just general historical methods, there are doubts as to whether Church history is adequately conceived of as a subfield or a field of specialization of historical studies (in German, *Bereichswissenschaft* or *Arbeitsbereich*; see, for example, Leppin 2014, 86–87; Beutel 1998, 5). The main reason for the doubts is that, in many epochs, secular and religious matters are so thoroughly interwoven that it is hard to see how to establish a boundary between secular and Church history. While this is certainly true, it is not a decisive reason to deny Church history the status of a historical subfield. Many historical subfields (such as history of medicine and history of biology) are considered to be well delineated even though they have vague boundaries and overlap with other subfields.

There is debate, however, as to whether Church history is *just that* – a specialized subfield of history delineated, however approximately, by its subject matter – or whether there is something more, which makes Church history a truly theological discipline.[22] Traditionally, Church history had to serve apologetic ends. The historian had to show that history (in the sense of the objective course of events) is in fact salvation history (*Heilsgeschichte*). In a long and painful process, Church history has been emancipated from this captivity. The second half of the 20th century saw a lively discussion among Catholic Church historians in Germany as to whether Church history should just assume history to be salvation history (Jedin 1962); should operate by the general methods of historical science except that, having reconstructed the historical facts, it then should concatenate and assess these facts from a particular religious perspective (Jedin 1970); should, at a minimum, allow itself to interpret historical findings in the light of faith (Iserloh 1970, Iserloh 1985); or should be kept strictly within the limits of general historical methods (Conzemius 1975). Nowadays, most Church historians will probably agree with Kurt Nowak's abstract characterization of Church history as »a historical discipline with a theological dimension.«[23] The theological dimension – or the »theological plus,« as I like to call it – then probably resides chiefly in the ends for which one undertakes the historical studies.[24] Church history studies are part of the self-reflection of a particular religious group, a self-reflection that is important for the group's identity and the continuity of its religious beliefs. Moreover, for a religious community – as can be drawn from the fact that it is almost always historians who work in the terrible field of abuse studies – historical studies play a role analogous to the role of memory in our personal lives, including reflection on and learning from our past successes and failures. If historical studies do indeed play such a reflective role for religious communities (something not all Church historians would necessarily agree about[25]),

---

22   The existence of this debate is mentioned briefly by Brox 2002, 7. Dassmann 1996, 9, holds that Church history as a theological discipline must »show God's salvific operation in the historical course of events« (my translation) and hence must presuppose theological premises that secular history would not dare to. See also Seeliger 1981, 14.

23   Nowak 2002, 464, my translation.

24   In the 19th-century debate, the »theological plus« was more evident in the evaluative dimension of historical studies: it was held that only the believer can »justly judge« historical facts concerning religiously meaningful institutions. See Scheidgen 1990, 218.

25   For a broad overview of the various positions with respect to the nature of Church history, see Baumann 2019, 59–69, and Leppin 2014. Holzem 2000 acknowledges a *memoria* role of Church history, although he maintains that the normative ideas ("Wertideen", Max Scheler) guiding Church history stem not only from the Church but from all kinds of discourses.

then an additional form of discovery is possible within Church history, namely a sort of self-discovery: finding out something about oneself by considering one's history.[26]

To sum up: In theology, there are historical discoveries. Historical discoveries concern historical facts as well as historical theories (or narratives). The discoveries on the side of theory concern relations among or within the theories and relations between theories and facts. If Church history serves the sort of reflective ecclesial purpose just described, then it also admits of discoveries analogous to learning from one's own history.

11.4   (Other) Pure and Impure Empirical Discoveries

In theology, we find discoveries not only of historical empirical facts but also of contemporary empirical facts.[27] In some theological disciplines, like pastoral theology or Christian social ethics, empirical knowledge plays an important role.[28] When Christian social ethics is concerned with a moral assessment of our social institutions (including forms of economic life) under the aspect of justice, its first task is to understand the relevant social phenomena. Modern social problems have become so complex that they cannot be properly grasped without the help of the social sciences, political science, economics, and other disciplines.[29] The same holds for pastoral theology. Recognizing and developing an understanding of all the practical forms of Christian life, individual as well as institutional, requires a firm grip on how human life, human organizations, and human societies actually function. So, these theological disciplines have enormous need of empirical knowledge.

On the whole, such empirical knowledge is acquired by the other disciplines I mentioned, such as social sciences and economics. Some of it, however, is gathered by theologians – especially data closely related to the Church or to Christianity. Such data includes statistics about weekly worship attendance, the declining number of priests, the trajectory of church finances in the form

---

26   Cf. Droysen's bon mot about general history: »History is humankind's becoming and being aware of itself« (Droysen 1882, 36, my translation), and Jedin's application to Church history: »Church history is the self-knowledge of the Church and of Christians« (Jedin 1970, 33, my translation).

27   To be sure, there is no clear and objective boundary between historical and contemporary facts. Where one draws the line depends on one's purposes.

28   See Möhring-Hesse 2019.

29   Kerber 1991, 172; see also Lienkamp 1995, 45–46, which frequently adverts to the relevant literature.

of taxes or donations, social studies of Catholic milieus in their diachronic development, survey results concerning what ordinary religious believers really believe, and similar types of information. Hence, theologians do generate empirical data.

Furthermore, such data calls for theoretical explanation. Real trends need to be carefully distinguished from seeming trends, probable causes of the developments need to be identified and carefully distinguished from seeming causes, and so forth. In short, once again we need theories based on empirical evidence – theories with the ordinary functions that empirical theories have in other scientific fields: explaining the data, allowing for the projection of future developments, informing the design of concrete measures, and guiding the evaluation of previously applied measures. (I will take up such procedural applications in the next section.)

While all these types of empirical data and theories are relevant for theology, and some of them are generated by theologians, these facts alone would probably not render the field where such knowledge is gathered a theological discipline. To be truly theological, the discipline must manifest something more – at least, a genuine theological perspective in assessing such knowledge.

Perhaps, a theological perspective can even generate its own kind of empirical knowledge. I will discuss the possible forms of such empirical knowledge using Christian Social Ethics (CSE) as an example. CSE reflects on the structures of social relations under the aspect of justice within the normative framework of Christian faith. It considers not only (as ethics and moral theology do) how social relations *should be* ordered, but also (as sociology does) how they *are* ordered and (as political science does) how they *can be* ordered. Hence, CSE includes a broad variety of possible discoveries. On the level of facts, to start with, there is a continuum of potential discoveries ranging from »pure« empirical facts about the existence of particular social structures to »impure« facts concerning the question whether such structures are just. Putative answers to such questions can be considered »impure facts« because, on the one hand, they depend on a normative conception of justice, which makes them »impure« (so that there may be enduring disagreement about them even when there is no disagreement about the »pure« facts),[30] while on the other hand, whether a structure does or does not conform to a certain normative

---

30   Moral realists would probably reject my talk of »impure facts« as, according to their position, true moral sentences are factual in just the same way as other true sentences. However, my argument requires only a distinction between two sorts of facts, normative and non-normative. What they are called does not affect the argument. One might call my pure facts »non-normative« and my impure facts »normative.«

conception also depends on how the world really is, which justifies calling the answers to such questions »facts.«

Besides the level of facts, discoveries are also possible at the level of theories. Here again, a distinction can be made between »pure« and »impure.« »Pure« tasks for a theory would include offering synchronic and diachronic understandings and explanations of social structures. »Impure« elements would include generalized normative assumptions as appear in theories of social justice. Such normative assumptions are the element most strongly influenced by religious beliefs. Examples of theory-level, or meta-level, discoveries in the literature of CSE include the discovery that in the light of current empirical knowledge about social life, abstract Christian norms can be concretized so as to lead to particular recommendations for action;[31] and the discovery by moral theologian Bruno Schüller in the 1970s that C. D. Broad's terminology can be used to classify traditional modes of norm justification in the Catholic theological tradition.[32]

It is thus evident that one can conceive of CSE as including empirical research, even though it is intimately connected to normative religious positions. The inclusion of particular normative conceptions from Christianity may be seen as the »theological plus« that demarcates CSE as a theological discipline from sociology and political science. Finally, many theologians conceive of CSE as also comprising reflections about the interrelation of social ethics and faith: How is a certain normative premise justified by the elaborated self-understanding of faith that theology continuously develops? Such reflections connect CSE deeply with other branches of systematic theology. What (non-empirical) discoveries are possible in systematic theology generally will be onstage in the section after next.

11.5   Action-oriented Discoveries

Let us return to the observation that there are non-historical theological subdisciplines that include empirical research, such as pastoral theology and CSE. Besides explaining and evaluating the world, these subdisciplines are also concerned with reflecting on possible ways to change the world. Such practical reflections involve discoveries of another kind that I call »action-oriented discoveries« – discoveries that concern the ways we act in the world.

---

31   Wiemeyer 2004, 141–142, 151.
32   Weiß 2019, 205.

Action-oriented discoveries are intimately connected to the empirical discoveries we have already discussed. For example, the available data on the trajectory of church taxes in some countries allows for forecasting the arc of tax revenue in the next five to ten years to some degree. Given that such a forecast reveals a severe decline in financial resources, the financial basis of certain charity programs may be found to be at risk. This finding may motivate devising measures to reduce or reorganize those programs or to find alternative funding for them. Ideally, such measures would also be evaluated after having been implemented.

While collecting and explaining data, producing forecasts, and evaluating implemented measures belong to the descriptive-explanatory enterprise, the task of developing new measures is of a special nature. It concerns not so much what is or will be the case as what we *can make* the case or even what we *intend to cause to be* the case by taking a given action. Just as political science may generate recommendations, this branch of research results in proposals for decision-makers (such as bishops, synods, or politicians) who are in a position to initiate corresponding courses of action. I hold that discoveries can be made in this sphere as well. Here, normative deliberations about what we want to change go hand-in-hand with instrumental assessments of what are the best means to reach the intended ends, considerations about the intended consequences and side effects of different types of measures, empirical findings about the effects and effectiveness of already implemented measures, and the like.

Discoveries can be made in these areas of consideration as well as in the complex interplay among them that influences the development of new measures. For example, consider the inventor of a new marketing instrument, a new public relations strategy, or a new concept of personnel resource allocation. If there is a clear-cut distinction between discovery and invention, such action-related discoveries may be considered to belong more to the invention side. Nevertheless, one can say that an inventor was successful in an epistemic undertaking (if concerning practical matters) and, to repeat, that he or she has *discovered that there is or can be* something like the invented measure.

In practice, descriptive and action-oriented elements are rarely separated but are almost always treated jointly. For example, if there is a decline in vocations for the priesthood, a pastoral theologian will first inquire into the reasons for the decline. In the light of such insights, she then may develop concrete proposals for action, such as improving the attractiveness of the Church as an employer or launching social media campaigns that target young people interested in religious topics. When a bishop decides to implement these measures, the pastoral theologian may also monitor the outcomes of the measures, differentiate – in case of a partial failure – between unsuitability of the measures and adverse conditions, and, finally, share her experiences with the scientific community, enabling the community to learn lessons that will benefit future practical projects of this kind. In such an ordinary interplay between an academic discipline such as pastoral theology and the practice of the Church, empirical and action-related discoveries are intertwined.

That the invention or discovery of practical measures may come about by accident or by trial and error does not in principle set these discoveries apart from other scientific discoveries. Granted, finding the right measures requires not only rational deliberation and calculation, but also experience, prudence, and wisdom. But the same is true of devising the right (formulations of the) laws of nature in the natural sciences.

In practical disciplines, the boundary between scientific and non-scientific discoveries is even vaguer than elsewhere. The activities of our pastoral theologian may be more or less scientific in nature. If she simply comes up with practical ideas or applies measures developed elsewhere, her discoveries will not appear particularly scientific. If, in contrast, she systematically develops a certain kind of measure to new lengths in the light of evaluations, combines it with other »modules« in intellectually sophisticated ways, and accompanies the implementation with theory-related projects such as simulations or data collection, then her discoveries will surely be regarded as scientific.

Under the heading of action-oriented discoveries, I also subsume ethical discoveries in the stricter sense involving normative reasoning. As ethics is treated in other contributions to this book, I will keep my considerations rather brief here. In my view, at least two kinds of discoveries are possible in ethics. First, there are empirical findings about our moral practice. These findings belong with the empirical discoveries we have already discussed. Second, there are findings about normative ethics. For example, an ethicist finds a new justification for some norm or develops a whole new framework for such justifications. (Think of Jeremy Bentham, who developed the first comprehensive version of utilitarianism.) Or a moral theologian finds a new way to make a traditional moral teaching of the Church plausible in the eyes of her contemporaries.[33] Because of such examples, I hold that it is indeed possible to make discoveries in ethics and moral theology. What exactly the nature of

---

33   A central task of theological ethics/moral theology is »the critical analysis of one's own
      tradition and the examination of its arguments« (Weiß 2019, 211, my translation). Ethical
      arguments in general can help »to confirm the validity of controversial norms, to improve
      their justification, to clarify their formulation, or to disprove their validity« (Weiß 2019,
      207, my translation).

such discoveries is, I happily leave to the experts in (meta-)ethics. My point with respect to theology is just that discoveries can be made in moral theology proper – in the theological discipline that not only inquires into morality but also is responsible in some way to the moral doctrine of the Church (not that all present-day moral theologians would follow me in attributing this ecclesiastical obligation to their discipline).

11.6 Hermeneutical or Doctrine-related Discoveries

Many discoveries in theology do not concern empirical or historical data in a narrow sense. They are, instead, interpretational or hermeneutical in nature. Their empirical basis, if there is one, consists in texts (or other forms of cultural transmission). But the texts do not simply play the role of historical evidence – although sometimes they do, as when a historian of theology studying the Church fathers discovers that one of them must have been influenced by another. The texts also constrain and inspire interpretations and meta-level insights. For this reason, there can be discoveries of a different kind, which bear much more resemblance to philosophical discoveries than to historical ones.

As I mentioned at the outset, the kinds of discoveries that are possible in systematic theology depend heavily on the assumed conception of (systematic) theology. The variety of conceptions on the market is great. They even differ as to systematic theology's function. Is it simply the logical development of an unchanging, eternal doctrine, as some traditionalists would have it? Is it the production of religious narratives that may powerfully influence a religious audience, but that consist in a free play of associations, appeals, expressions of affectedness, and the like, as some postmodernists would have it? Is it just about understanding the historical development of certain doctrines, as some historically inclined theologians maintain? This enumeration of disparate conceptions of systematic theology could go on and on. The point is that which conception of systematic theology one pursues has a profound impact on whether one can countenance discoveries in the field and, if so, what their nature might be.

I have argued elsewhere[34] that if systematic theology is to be a scientific endeavor, it is best construed in analogy to the natural sciences with their interplay between theory and data. On this construal, systematic theology works in a parallel way with religious theories and »data.« Religious theories

---

34   See my 2018.

are something like the objective counterpart to the subjective religious belief systems of persons. One might call such theories »doctrines,« but I prefer to reserve that term for the official teachings of a religious community such as a church. A religious theory is (intended to be) a coherent body of sentences, some of which are religious in content. Religious theories are subject to criteria like coherence, consistency, and adequacy with respect to the data. The role of the data is played by the religious propositions of a religious community. Such religious propositions must be distilled in a complex hermeneutical process from the whole body of the religious tradition, on the one hand, and the world-view of the hermeneutical subject, on the other hand. The distillation process must take into consideration the dependence of certain teachings on further assumptions, the historical development of these teachings, how binding they each are, and many other factors.

According to this raw picture, systematic theology develops a plurality of legitimate religious theories. This plurality is an offering to the religious indi-vidual and to the religious institution (not only for catechesis and preaching but also for further development of the teaching based on a deepened self-understanding). Why should there not be many different religious theories, each constituting a different way to bridge the gap between personal convic-tions and the shared identity of a religious group?

On the assumption that this picture is more or less adequate, a whole array of types of scientific discoveries in systematic theology is possible. These types of discoveries include:

1.   Discovering that a proposition belongs to the religious beliefs of the Church (with a certain degree of bindingness, certain dependence rela-tions to other propositions, and so forth).

2.   Discovering that certain parts of a religious theory are in tension with others or even seem to contradict them (consistency and coherence problems).

3.   Discovering that such problems can be solved by distinguishing between different meanings of the terms involved.

4.   Discovering that such problems can be solved by reinterpreting a certain part of the theory.

5.   Discovering that such problems can be solved by historical relativization: that certain parts of a former theory may be judged accidental and of minor importance.

6.   Discovering that a certain part of a religious theory can be better connected to other parts by certain arguments (inner coherence improvements).

7.   Discovering that a certain part of a religious theory can be better adapted to external knowledge (external coherence improvements).

8. Discovering that there is potential to incorporate other theories on offer in, for example, philosophy, psychology, ethics, or the social sciences.

11.7 Genuinely Theological Discoveries

This paper has set out to bring to light the existence of a great variety of scientific discoveries in the subdisciplines of theology. As those subdisciplines are in many cases intimately connected with their secular counterparts (e.g., Church history with historical studies, exegesis with philology and literary studies), the question arises whether there are discoveries that require theology as a whole and would not be made if only the subdisciplines existed as areas of specialization within the secular counterpart disciplines. I will term such discoveries »genuinely theological.« Are there genuinely theological discoveries?

This question may take on a more anthropological or sociological aspect: Do people working in the theological subdisciplines perceive themselves more as theologians with a specialization in a given theological subdiscipline, or more as members of the respective secular counterpart disciplines with a specialization in Church- or religion-related topics? Does a Church historian consider herself a theologian specializing in historical studies or a historian specializing in Church or religion? According to my experience, there is a continuum between the two poles. Most of my colleagues in theology departments would see themselves in both ways, with varying emphasis on one or the other side.

The question concerning genuinely theological discoveries is not so easy to answer. The area that was most challenging with respect to whether there are discoveries at all, namely systematic theology, is the easiest to deal with in this respect: If we have agreed that there are scientific discoveries in the area of systematic theology proper, then we will most probably be in accord with the view that such discoveries are genuinely theological discoveries. But what about discoveries in the subdisciplines that have secular counterparts?

I have suggested that there might be a »theological plus« in theological subdisciplines that distinguishes them from their secular siblings. If there is such a »theological plus,« then there is something more to the discoveries than just belonging to a Church- or religion-related area of specialization within the secular counterpart discipline. However, this argument is weak insofar as I have suggested that the »theological plus« is primarily grounded in the theological ends the subdiscipline pursues. Does a religious motivation or the use for some religious ends make a historical insight so special as to justify distinguishing it from other historical insights not so motivated or applied? This question suggests that the argument from the »theological plus« does not definitely discredit the suspicion that there are no genuinely theological discoveries in those subdisciplines.

A stronger case for genuinely theological insights can be made by considering insights that arise from a combination of different subdisciplines. As an example, consider St. Paul's odd commandment that women should remain silent in the assembly (1 Cor 14, 33–34). Here exegesis, if it were just a philological business, could not do much more than discover different possible senses of the text, or call into question whether this commandment is really from St. Paul. Historical insights into the relatively weak position of women in ancient oriental societies could contribute reasons for relativizing such a »command.« Much more power, however, is evinced when centuries of exegetical, historical, practical, and systematic research culminate in a firm theological consensus that all human beings have the same dignity and that all members of a religious community are created equal. If the »theological plus« of exegesis implies that such a general theological consensus, which may itself have been influenced by exegesis, should now in turn place hermeneutical constraints on future exegesis, then such a consensus can influence exegetical results and thereby help to remedy long-established injustices. When, in ways like this, combined research from multiple theological subdisciplines leads to a consensus that is accepted as background knowledge for future research, that consensus and the results that flow from it are examples of genuinely theological discoveries.

11.8   Conclusions

This paper has presented an idealized, but also to some degree realistic, picture of theology. Theology consists of a cluster of subdisciplines, many of which have secular counterparts with which they share most of their methods. Hence, theology admits of at least such discoveries as would be recognized in the secular counterpart disciplines. I tried to identify what might count as a »theological plus« that distinguishes a theological subdiscipline from its secular counterpart, and I indicated that there is controversy regarding whether one should accept such a »theological plus« and what it consists in. In the end, I think, there are good reasons to accept such a »theological plus« in one or another form. A promising task for future research is to investigate the miscellaneous conceptions of what constitutes such a »plus.« The proposed concept of »impure empirical facts« – facts that depend both on how the world is and on certain normative presuppositions – offered one way to conceive of a

»theological plus.« The example of a theological consensus influencing exegesis offered another way.

Theological discoveries can be classified into four groups: historical discoveries, other empirical discoveries, action-based discoveries, and interpretation-based discoveries. In each group, the nature of the discoveries depends for the most part on the nature of the methods used. The unique case of systematic theology called for special attention, both because conceptions of systematic theology vary greatly and because it does not feature the immediate relation to »facts« or »data« that more historical-empirical subdisciplines do. I proposed understanding systematic theology as working with religious theories. The adequacy conditions on these theories are set – in analogy with the role of data in the natural sciences – by single religious propositions that must be distilled from a religious tradition in a complex hermeneutical process. This understanding of systematic theology gave rise to an extensive taxonomy of possible discoveries within that subdiscipline.

References

Baumann, N. 2019. »Verortet in Zwischenräumen? Alte Kirchengeschichte und Patrologie als wissenschaftliche Disziplinen.« In *Die Wissenschaftlichkeit der Theologie*, vol. 2, edited by B.P. Göcke & L.V. Ohler, 53–84. Münster: Aschendorff.

Beutel, A. 1998. »Vom Nutzen und Nachteil der Kirchengeschichte. Begriff und Funktion einer theologischen Kerndisziplin.« In *Protestantische Konkretionen: Studien zur Kirchengeschichte*, edited by A. Beutel, 1–27. Tübingen: Mohr Siebeck.

Bochenski, J. 1965. *The Logic of Religion*. New York: New York University Press.

Brox, N. 2002. *Kirchengeschichte des Altertums*. 6. ed. Düsseldorf: Patmos.

Buntfuß, M. & Fritz, M. 2014. *Fremde unter einem Dach? Die theologischen Fächerkulturen in enzyklopädischer Perspektive*. Berlin: De Gruyter.

Conzemius, V. 1975. »Kirchengeschichte als ›nichttheologische Disziplin‹. Thesen zu einer wissenschaftstheoretischen Standortbestimmung.« *Theologische Quartalschrift* 155: 187–197.

Dassmann, E. 1996. *Kirchengeschichte II/1*. Stuttgart: Kohlhammer.

Droysen, G. 1882. *Grundriss der Historik*. 3rd ed. Leipzig: Veit 1882.

Gatzemeier, M. 1974–1975. *Theologie als Wissenschaft?* 2 volumes. Stuttgart: Frommann-Holzboog.

Göcke, B. et al., eds. 2018–2019. *Die Wissenschaftlichkeit der Theologie*. Volumes 1–3 (=STEP 13/1–3). Münster: Aschendorff.

Holzem, A. 2000. "Die Geschichte des 'geglaubten Gottes'. Kirchengeschichte zwischen 'Memoria' und 'Historie'". In *Katholische Theologie studieren: Themenfelder und Disziplinen*, edited by A. Leinhäupl-Wilke & M. Striet, 73–103. Münster: LIT.

Iserloh, E. 1970. »Was ist Kirchengeschichte?« In *Kirchengeschichte heute: Geschichtswissenschaft oder Theologie?*, edited by R. Kottje, 10–32. Trier: Paulinus.

Iserloh, E. 1985. »Kirchengeschichte – Eine theologische Wissenschaft.« *Römische Quartalsschrift für christliche Altertumskunde und Kirchengeschichte* 80: 5–30.

Jedin, H. 1962. »Einleitung in die Kirchengeschichte.« In *Handbuch der Kirchengeschichte*, vol. 1, edited by H. Jedin, 1–68, Freiburg: Herder.

Jedin, H. 1970. »Kirchengeschichte ist Theologie und Geschichte.« In *Kirchengeschichte heute: Geschichtswissenschaft oder Theologie?*, edited by R. Kottje, 33–48. Trier: Paulinus.

Leppin, V. 2014. »Die Kirchengeschichte im Kreis der theologischen Fächer. Historische Offenlegung der vielfältigen Möglichkeiten christlicher Religion.« In *Fremde unter einem Dach? Die theologischen Fächerkulturen in enzyklopädischer Perspektive*, edited by M. Buntfuß & M. Fritz, 69–93. Berlin: De Gruyter.

Lienkamp, A. 1995. »Quellen der Ethik? Zur erkenntnistheoretischen Bedeutung der Sozialwissenschaften für die Soziallehre der Kirche.« In *Brennpunkt Sozialethik: Theorien, Aufgaben, Methoden*, edited by M. Heimbach-Steins, A. Lienkamp, & J. Wiemeyer, 45–68. Freiburg: Herder.

Michel, J.G. 2020. »Could Machines Replace Human Scientists? Digitalization and Scientific Discoveries.« In *Artificial Intelligence: Reflections in Philosophy, Theology, and the Social Sciences*, edited by B.P. Göcke & A. Rosenthal-von der Pütten, 361–376. Paderborn: mentis/Brill.

Möhring-Hesse, M. 2019. »Wissenschaftlichkeit der theologischen Sozialethik.« In *Die Wissenschaftlichkeit der Theologie*, vol. 2, edited by B.P. Göcke & L.V. Ohler, 217–243. Münster: Aschendorff.

Morscher, E. 1974. »Das Basis-Problem der Theologie.« In *Der Modernismus: Beiträge zu seiner Erforschung*, edited by E. Weinzierl, 331–368. Graz: Styria.

Morscher, E. 2003. »Was ist gute Theologie?« In *Was ist gute Theologie?*, edited by C. Sedmak, 324–333. Innsbruck: Tyrolia.

Murphy, N. 1990. *Theology in the Age of Scientific Reasoning*. Ithaca: Cornell University Press.

Nowak, K. 2002. »Wie theologisch ist die Kirchengeschichte?« In *Kirchliche Zeitgeschichte interdisziplinär: Beiträge 1984–2001*, edited by J.-C. Kaiser, 464–473. Stuttgart: Kohlhammer.

Pannenberg, W. 1973. *Wissenschaftstheorie und Theologie*. Frankfurt: Suhrkamp.

Peukert, H. 1978. *Wissenschaftstheorie, Handlungstheorie, fundamentale Theologie*. Frankfurt: Suhrkamp.

Scheidgen, H. J. 1990. »Mittlere und neuere Kirchengeschichte.« In *Katholische Theologie heute*, edited by J. Wohlmuth, 216–228. Würzburg: Echter.

Scholz, H. 1931. »Wie ist eine evangelische Theologie als Wissenschaft möglich?« *Zwischen den Zeiten* 9: 8–53. Reprinted in *Theologie als Wissenschaft: Neudrucke und Berichte aus dem 20. Jahrhundert*, vol. 43, edited by G. Sauter, 221–264, München: Kaiser, 1971.

Seckler, M. 1988. »Theologie als Glaubenswissenschaft.« In *Handbuch der Fundamentaltheologie. Vol. 4: Traktat Theologische Erkenntnislehre*, edited by W. Kern, H.J. Pottmeyer, & M. Seckler, 180–241. Freiburg: Herder.

Seeliger, H.R. 1981. *Kirchengeschichte – Geschichtstheologie – Geschichtswissenschaften. Analysen zur Wissenschaftstheorie und Theologie der katholischen Kirchengeschichtsschreibung.* Düsseldorf: Patmos.

Tapp, C. 2015. »Kann es im kirchlichen Glauben Neues geben?« In *Die Theologie und »das Neue«: Perspektiven zum kreativen Zusammenhang von Innovation und Tradition*, edited by W. Damberg & M. Sellmann, 139–168. Freiburg: Herder.

Tapp, C. 2017. »Logik in Religionsphilosophie und Theologie.« In *Logik in den Wissenschaften*, edited by P. Klimczak & T. Zoglauer, 83–107. Paderborn: Mentis.

Tapp, C. 2018. »Wissenschaftliche Theologie: Anforderungen und Grundlinien eines theorie-orientierten Modells.« In *Die Wissenschaftlichkeit der Theologie*, vol. 1, edited by B.P. Göcke, 203–225. Münster: Aschendorff.

Tetens, H. 2015. »Der Gott der Philosophen. Überlegungen zur Natürlichen Theologie.« *Neue Zeitschrift für Systematische Theologie und Religionsphilosophie* 57 (1): 1–13.

Weingartner, P. 1971. *Wissenschaftstheorie: Einführung in die Hauptprobleme.* Stuttgart: Frommann-Holzboog.

Weiß, A.M. 2019. »Vernünftig argumentieren in der Theologischen Ethik.« In *Das Theologische in der Theologie: Wissenschaftstheoretische Reflexionen – methodische Bestimmungen – disziplinäre Konkretionen*, edited by F. Gmainer-Pranzl & G.M. Hoff, 203–214. Innsbruck: Tyrolia.

Wendel, S. 2002. »Die Rationalität der Rede von Gott. Thesen zur Legitimation der Theologie als Wissenschaft.« *Stimmen der Zeit* 220 (4): 254–262.

Werbick, J. 2010. *Einführung in die theologische Wissenschaftslehre.* Freiburg: Herder.

Wolf, H. & Seiler J. 2004. »Kirchen- und Religionsgeschichte.« In *Aufriß der historischen Wissenschaften in sieben Bänden, Bd. 3: Sektoren*, edited by M. Maurer, 271–338. Stuttgart: Reclam.

Bullshit in Science? – On Epistemic Norms, Credibility and the Role of Science in Society

Susanne Hahn

The starting point of my essay[1] is the fact that scientific discoveries are established as *scientific* discoveries in a normative system of science that owes its credibility in large part to its freedom from external intervention.[2] In the following I want to reflect on a connection between the business of generating knowledge, more specifically: doing science, on the one hand, and, on the other hand, the *institutional framework* of knowledge production as it is set, for example, in Germany by Article 5 of the constitution.[3] Ultimately, this is a question of linking epistemological action and political action. In the background there is an assumption that belongs to the legacy of political philosophy in the footsteps of the French Revolution: Liberal democracies rely on political self-determination mediated by representatives. In order to be able to perform these tasks in a meaningful way, both citizens as voters and representatives as decision-makers need knowledge about the subject areas in which they can intervene by means of political measures. The delivery of these findings by science provides the basis for enlightened judgements.

John Stuart Mill has expressed concisely the connection between representative democracy and the prerequisites among voters and elected officials as follows:

> »Government consists of acts done by human beings; and if the agents, or those who choose the agents, or those to whom the agents are responsible, or the lookers-on whose opinion ought to influence and check all these, are mere masses of ignorance, stupidity, and baleful prejudice, every operation of government will go wrong; while, in proportion as the men rise above this standard, so will the government improve in quality; up to the point of excellence, attainable but nowhere attained, where the officers of government, themselves persons of

---

1 The author thanks Jan Michel, Geo Siegwart and Christian Tapp for helpful comments.

2 The normativity of scientific research mentioned above is a link between the issue of scientific discoveries and the topic of my essay; see Michel 2019.

3 Article 5 (3) of the Basic Law for the Federal Republic of Germany reads: »Kunst und Wissenschaft, Forschung und Lehre sind frei. Die Freiheit der Lehre entbindet nicht von der Treue zur Verfassung.« – »Arts and sciences, research and teaching shall be free. The freedom of teaching shall not release any person from allegiance to the constitution« (https://www. gesetze-im-internet.de/englisch_gg/englisch_gg.html#p0034).

superior virtue and intellect, are surrounded by the atmosphere of a virtuous and enlightened public opinion.

The first element of good government, therefore, being the virtue and intelligence of the human beings composing the community, the most important point of excellence which any form of government can possess is to promote the virtue and intelligence of the people themselves.« (Mill [1861] 1977, 390)

Science, however, has come under pressure from various sides in recent years. In my essay I would like to develop a perspective on the *production of truths* or weaker forms of cognition, e.g. conjectures, *as norm-guided action*, which is institutionally granted a free space due to its own system of norms. Scientific action that disregards these norms endangers this freedom. And in another direction: Political action that attacks the freedom and independence of the scientific norm system endangers the knowledge base for deliberative democracy.

In order to create such a perspective framework, I would like to address the following issues: In a first step I develop a view of the generation of knowledge as norm-guided action (1). In the second section I present Harry Frankfurt's very stimulating idea to label and classify the phenomenon of epistemologically improper behaviour with the term »bullshit« (2). On this basis, I sketch three examples of such misconduct by scientists as actions that violate scientific norms (3). The last section reflects on the role of science in liberal democracies and the duties that arise from it to ensure the credibility of science (4).

12.1   Acting to Generate Knowledge as Norm-guided Action

I presuppose that in the context of discussing scientific discoveries the importance of knowledge and the aim of truth is uncontroversial. Beside an intrinsic interest in truth in scientific discourse we have an essential interest in truths in everyday matters. We want to act successfully, we want to realise certain states, achieve goals and the success of these concerns depends on our being able to rely on sound knowledge. As Moritz Schlick already put it in 1918:

> »That all knowledge originally served only practical purposes is an indubitable, much emphasized truth.« (Schlick [1918] 1974, 95)

Or, more recently, in Harry Frankfurt's formulation:

> »We really cannot live without truth. We need truth not only in order to understand how to live well but in order to know how to survive at all.« (Frankfurt 2006, 36)

In the case of a stroke, for example, it is of vital importance to know whether it is the result of a vascular occlusion – this is the case in about 80% of cases – or the result of a cerebral haemorrhage. The standard immediate therapy for occlusion is the application of a drug that eliminates it. This measure, however, is associated with a bleeding risk that would be fatal if it were a stroke caused by cerebral hemorrhage. It is therefore important to determine the cause of the stroke as accurately as possible. For cases like these, an appropriate procedure exists: The exclusion of a brain hemorrhage can be achieved by computer tomography. Action in this situation depends largely on the diagnosis on which it is based – the established knowledge can play a vital role here. In the case presented, the securing of knowledge is device-supported: If a computer tomography of the head produces a specified image, then one is allowed to perform a strong act of knowledge, namely one can make a *statement*: There is no cerebral hemorrhage. If one is allowed to *state* a proposition, then this proposition is correctly presented as true.[4] If, by contrast, the computer tomography delivers a different picture and the examiner nonetheless *states* the proposition »There is no cerebral hemorrhage« and thus presents it as true, the proposition is *incorrectly* stated as true. The utterer of this sentence acts against the underlying norm for stating. We would sue him for a medical error and not accept if he pointed out that there were »alternative facts«[5]. – Gaining knowledge is norm-guided action.

However, the situation is not always so simple – presumably it is even the case that in many interesting situations in which we are dependent on knowledge for our actions no true propositions are gained, i.e. no statements or other *strong* acts of knowledge can be made, but only *weaker* acts such as conjectures or estimations. This insight has essential consequences for the handling of the results of the business of knowledge.

Medical tests, such as the PSA test in prostate cancer screening or the test for hidden blood in the stool, do not entitle to make statements – i.e. to utter a strong affirmative cognitive act – of the type »there is (no) a prostate cancer«

---

4   The speech act of stating is one of several strong affirmative cognitive acts. Other strong affirmative acts are asserting, setting-as-postulate, defining etc. For the underlying epistemological and especially alethiogical framework that displays a thoroughly pragmatic understanding of generating knowledge by bringing together a use theory of meaning and speech act theory see Siegwart 2007, especially 48. – The norms mentioned here are epistemic norms that guide the respective cognitive act, they are constitutive of the concrete insight. They differ from the more general norms Robert K. Merton identifies in the scientific ethos which »shape a system of communication« in the scientific community (Weingart 2015, 11).

5   This now famous phrase was chosen by Kellyanne Conway, Donald Trump's advisor, to justify the obviously false statement by Sean Spicer, then White House Press Secretary, that the inauguration of Donald Trump is the one with the most visitors ever.

or »there is no intestinal cancer«, but only to make strong conjectures, usually in combination with other diagnostic characteristics. These provisos, i.e. these considerations concerning the strength of the act of cognition, which can be carried out on the basis of underlying norms, become all the more apparent when dealing with findings from the social sciences: Statements like the one that the new generation is no longer interested in salary levels or in typical status symbols such as company cars etc. are not only undifferentiated, but also fall back on a small sample size. As this very rough sketch shows: The production of truths or weaker forms of cognition is guided by standards and different degrees of evidence entitle to different cognitive acts.[6]

12.2   Dealing with Insights: Bullshit as a Problem

In his essay *On Bullshit*, Harry Frankfurt characterises a behaviour that threatens the orientation towards truth as the basis for action. I tend to agree with Frankfurt's diagnosis that the loss of confidence in politics, the media and science, which I would like to discuss in the next part, is not so much due to the liars as to a widespread phenomenon which Frankfurt describes using the term »bullshit«. Bullshit differs from lying:

> »Telling a lie is an act with a sharp focus. It is designed to insert a particular falsehood at a specific point in a set or system of beliefs, in order to avoid the consequences of having that point occupied by the truth. This requires a degree of craftmanship, in which the teller of the lie submits to objective constraints imposed by what he takes to be the truth. The liar is inescapably concerned with truth-values. In order to invent a lie at all, he must think he knows what is true. And in order to invent an effective lie, he must design his falsehood under the guidance of truth.« (Frankfurt 2005, 85f.)[7]

Bullshit is a different story: According to Frankfurt, the essence of this kind of announcement is that the person making the statement *doesn't care whether what they say is true or false.*

---

6   For the respective cognitive acts and the norms that support them see Siegwart 2007, 48–50.

7   Following Wolfgang Künne's exposition on lying, the following characterization can be given: x lies iff x presents a statement A as true and x is convinced that A is false and x has the intention to make someone y believe that A is true by putting A in the »as true« position. See Künne 1985, 119.

> »It is just this lack of connection to a concern with truth – this *indifference to how things really are* – that I regard as of the essence of bullshit.« (Frankfurt 2005, 59, emphasis S.H.)

Based on Frankfurt's distinction between lying and bullshitting, I distinguish two types of bullshitting: strategic unscrupulous bullshitting and careless bullshitting.

Persons who strategically and unscrupulously bullshit are indeed indifferent to truth and falsehood insofar as they do not set them as standards for their utterances. But unscrupulous bullshitters are not indifferent in principle, they use their utterances deliberately to achieve their aims and want to deceive, but in a certain way: This is what I call »strategic bullshitting«.

> »The fact about himself that the bullshitter hides, on the other hand, is that the truth-values of his statements are of no central interest to him; what we are not to understand is that his intention is neither to report the truth nor to conceal it. [...] His eye is not on the facts at all, as the eyes of the honest man and the liar are, except insofar they may be pertinent in his interest to getting away with what he says. He does not care whether the things he says describe reality correctly. He just picks them out, or makes them up, to suit his purpose.« (Frankfurt 2005, 91f.)

We see another type of bullshitting when people speak about matters of which they have no clue. This is what I call »careless bullshitting«.

> »Bullshit is unavoidable whenever circumstances require someone to talk without knowing what he is talking about. Thus the production of bullshit is stimulated whenever a person's obligations or opportunities to speak about some topic *exceed his knowledge* of the facts that are relevant to that topic. This discrepancy is common in public life, where people are frequently impelled – whether by their own propensities or by the demands of others – to speak extensively about matters of which they are to some degree ignorant. Closely related instances arise from the widespread conviction that it is the responsibility of a citizen in a democracy to have opinions about everything, or at least everything that pertains to the conduct of his country's affairs. The lack of any significant connection between a person's opinions and his apprehension of reality will be even more severe, needless to say, for someone who believes it is his responsibility, as a conscientious moral agent, to evaluate events and conditions in all parts of the world.« (Frankfurt 2005, 103–106, emphasis S.H.)

While strategic bullshitters ruthlessly disseminate messages in the service of their objectives, this motivation is irrelevant to careless bullshitters. Instead, it is carelessness, lack of diligence, perhaps even presumption or a mixture thereof. Both types of bullshitting contribute to an erosion of credibility or trust in the public debate. In particular, the second type, careless bullshitting, poses a threat for the credibility of science and scientists. It can easily happen that scientists, especially in the context of public media presentations, make statements whose status of truth is not covered by the corresponding evidence or which go beyond their respective expertise. As presumably many readers can tell from their own experiences, this does not happen very often in the sense of strategic bullshitting. It is more likely to be fostered by the situational contexts in which the statements are made, e.g. in interviews, and it is supported by a supposed obligation the interviewed scientist feels to answer every question – instead of answering only those questions that can be answered on the scientist's expertise.

On the whole, one can try to classify behaviour in such a way that the orientation towards truth as honourable is guiding: There is honourable conduct and dishonourable conduct. Lying and bullshitting belong to the dishonourable conduct. With respect to bullshitting can distinguish between strategic and careless bullshitting.

12.3   Bullshit in Science (Communication) – the Violations of Norms

The German surveys called »Wissenschaftsbarometer« (»Science Barometers«), which assess the population's attitudes to science and research, indicate relative stability with respect to trust in science and research.[8] It should nevertheless be noted that slightly more than half of the respondents answered *undecided* (46%) or *negative* (7%) when it comes to trust in science and research. – Besides, in current political discussions there are also groups that question the informative and interpretative power of media and science equally. The catchiest buzzword is »lying press«. Fake news accusations, however, also belong here, as well. One could respond to these accusations by pointing to the fact that they come from a certain political corner, that they are not serious and that they are, therefore, not worth to be taken seriously and further analysed. But that seems too simple – they are accusations that are carried on in the population and express themselves in attitudes such as »those up there« as wells as in a criticism of the so-called elites and can ultimately flow into election decisions. Assume these accusations are not ignored or rejected immediately, but taken seriously and subjected to an examination: What diagnoses can be made about the state of the acquisition

---

8 https://www.wissenschaft-im-dialog.de/projekte/wissenschaftsbarometer/
   wissenschaftsbarometer-2019/.

and transmission of knowledge? What objectives are pursued with the science system in our society, in our state, and what norms and duties can be derived from these objectives?

Three examples shall serve as a basis for the reflection on misconduct of scientists:

12.3.1 *The Reassuring Seismologists*

In 2009, a devastating earthquake occurred in the city of L'Aquila, Italy, killing over 300 people. As a result, the city was initially no longer habitable. The interesting fact in the context of epistemic norms and the credibility of science is that six scientists were sentenced to long prison terms in 2012. They were members of the High Risk Commission, an independent advisory board to the Italian government. The seismologists were accused of providing the population with »imprecise, incomplete and contradictory« information concerning the earthquake risk.[9]

> »The experts travelled to L'Aquila one week before the earthquake and *downplayed* the earthquake risk in front of running television cameras. Commissioner Bernardo De Bernardinis' statements on the low risk are said to have caused at least 29 of the total 308 victims to remain in their homes, while those who spent the night on 6 April in the open despite reassurances survived the earth tremor.«[10]

In the alleged case, the convicted scientists show misconduct in at least two respects: They did not point out the probability of error in their prognosis, but made a general attempt to appease existing fears. In addition, they did not try to establish connections between the residual risk, the construction of the houses and the indicated actions. This would have been a contribution to instruction in dealing with probabilities. Overall, the seismologists carried out epistemological and recommendational actions that were not covered by the standards for the respective speech acts of stating and recommending.[11]

---

9 Schurz discusses the case with respect to value-neutral science, see Schurz 2013, 326–328. He criticises that the seismologists did not try to explain the probabilities of earthquakes explain to the inhabitants in order to let them decide what to do on their own.

10 Neue Zürcher Zeitung, 24 October 2012, https://www.nzz.ch/verurteilung-von-seismologen-loest-weltweit-reaktionen-aus-1.17715150, translation and emphasis S.H.

11 It is a different question whether this behaviour is sufficient for a criminal sanction. – In fact, the scientists were acquitted in the last court. See Staffler 2017.

12.3.2   *The Battle Over Red Meat*

Scientific findings on nutrition have received special attention in the last decades. Recently, a network of researchers has criticised that dietary recommendations are not sufficiently supported by the research evidence. The researchers mention the result of a large randomized control study stating that a 20% reduction in the consumption of red meat, contrary to the position and respective recommendation of the World Health Organization, has no effect on several cancer types.[12] The authors criticise that the suggestions to reduce the consumption of red meat rely on observational studies and therefore cannot meet the standards of randomized control studies. The article has provoked many negative reactions among experts.[13] Some of them insist on the results of the epidemiological studies and consider the randomised trial as not sufficiently relevant. There are also comments which emphasize that the authors only repeat what is already known. Currently there is no consensus, so that a reduced consumption of red meat is now recommended by some and not recommended by others. This is – as it seems to be usual when it comes to nutrition – confusing. However, some factors that account for disagreement with respect to nutritional questions can be identified. What happens is that scientists who regard red meat consumption as harmful often implicitly mix the direct physiological consequences for the human organism with animal welfare and greenhouse gas emissions from agriculture. Apart from that, nutritional studies face structural methodological problems as they cannot meet the conditions of controlled clinical trials; in this respect, strong affirmative speech acts, e.g. stating (see section 1), with respect to the effects of nutrition on the organism are not covered by the respective rules for stating and should be avoided. Finally, it can be noted that the scientists make recommendations. In order to do this correctly, scientists have to observe standards for correct scientific recommendations. A *correct* scientific recommendation would assume preferences, refer to the observed purpose-means relationships and make hypothetical recommendations. In nutrition studies, however, incorrect recommendations of two kinds are often found: Scientists take their own preferences as a basis without indicating it and make unconditional recommendations. Or they assume any preferences without identifying them explicitly and make recommendations.[14]

---

12   Johnston et al. 2018.

13   For a compilation see https://www.sciencemediacentre.org/expert-reaction-to-new-papers-looking-at-red-and-processed-meat-consumption-and-health/.

14   Johnston et al. 2018.

12.3.3 *Ethicists as Experts*

Many innovations have profound consequences for individual and collective lifestyles. In recent decades, the practice has emerged to accompany political regulations in areas such as the so-called life sciences or currently artificial intelligence or data management by way of expert commissions that often carry the word »ethics« in their titles, such as »Data Ethics Commission«.[15] Ethics experts make recommendations in such committees and incorporate them into guidelines. In these cases, the situation becomes even more complicated than with recommendations made by experts from the empirical sciences, as the standards of correctness for such prescriptive utterances are controversial or even unclear. This means that even the conditional character of regulatory recommendations is not consensual among practical philosophers: It is precisely the inclusion of interests in ethical considerations that is considered suspicious by some moral philosophical positions. »Real moral norms« are – in the Kantian tradition – independent of human interests. Some ethicists regard themselves as experts in morality and values and do not take into consideration that those affected by regulatory decisions may favour other values. The right to informational self-determination or digital self-determination may be an example for a discrepancy in the evaluation of experts and affected citizens. While the experts in the commissions emphasize the right to informational self-determination and thus view many possible applications, even in clinical contexts, rather critically, users – even after years of public education about the associated risks – still pass on their data in order to get access to certain apps or platforms. – The status of the statements of ethicists can thus always only be assessed relative to assumed, but not consensual, standards of correctness.

What do these examples show? First of all, I would like to assure you that I am not presenting an empirical study with a quantitative claim, but would rather like to develop a perspective for the role of science in public discourse. – Even though these examples given are individual cases, the mere fact that one does not have to search for such examples intensively provides a good starting point for my assumption that the outlined cases of misconduct are no exceptions. These examples, in which scientists perform speech acts that are not sufficiently supported by their research results, fuel scepticism towards science and contribute to a loss of credibility. Depending on the intentions of the scientists,

---

15 A recent article shows the flourishing of ethics committees and the formulation of guidelines by analysing 22 ethical guidelines for dealing with artificial intelligence. See Hagendorff 2020.

these are cases of either careless or strategic bullshitting: if scientists pursue their own goals, e.g. if they want to reduce other people's meat consumption, we are dealing with cases of strategic bullshitting. If, however, the scientists' statements are made under time pressure or out of a sense of duty to answer every question, even though they do not have the necessary expertise, we are dealing with cases of careless bullshitting. Unlike real deceptions, i.e. lies as they appear, when research data are falsified or when other people's results are reported as their own, these types of misconduct are less spectacular and less sanctioned, but contribute to undermining credibility nevertheless. Scientists who, today, promote at least five meals a day and the consumption of milk, and who, tomorrow, recommend at most daily two meals and the avoidance of milk, should not be surprised about a certain sceptic attitude against science. – The exceptional case of the seismologists suggests to think about a fundamental positioning of science with respect to the standards of the science system and the role this system shall fill in democratic societies (see section 4).

In addition, it would be desirable to have empirical investigations as to which framework conditions promote or impede such misconduct. The examples presented could provide indications of such conditions. They are characterised by the fact that the actions of the scientists take place in a public context – and not in a purely internal scientific context. It seems to me that two factors in particular are at work here: On the one hand, politicians call for clear recommendations for action. On the other hand, there is the pressure from the media to present results in extreme accentuation. While the former may be explained by politicians' desire to shift responsibility for inconvenient measures onto scientists, the pressure exerted by the media can be explained by the passing on of their own pressure. Under the conditions of online journalism and competition from social media, traditional media are confronted with the need to maximise the time users spent on their online versions. However, attention is usually not attracted by differentiated and careful presentations but by exaggerations and striking headings. Contacts with scientists also take place under this pressure. But scientists also voluntarily put themselves under pressure: one could ask what science slams organised by scientists contribute to the communication of scientific results and the nature of scientific work. Do they present a correct picture of scientific work? Or is it rather a distorted picture that depicts precisely those elements of science that are appealing to the public, entertaining and can be presented in three minutes? – If one combines these elements in a pessimistic way, one arrives at a scenario in which the exaggerated call for attention among all agents leads to the universal dissemination of the phenomenon described above as bullshit. Credibility in the systems of science and media would then fall by the wayside.

12.4 Credibility in the Interplay between the Right to Freedom and the System of Norms

Credibility and reliability are central categories for both political and scientific action. For science, the orientation towards truth and thus towards the norms of gaining knowledge in its own field of knowledge is essential. Violations of norms such as those described in the examples above are not deliberate deceptions. (As I already mentioned there are also real deceptions, e.g. when experimental results are falsified or when researchers pretend that other people's work is their own.) Unlike the case of the seismologists, the violation of epistemic norms is usually not sanctioned. This kind of careless bullshitting may therefore spread more widely and threaten the credibility of science as a whole. Nevertheless, it should be noted that this finding of non-compliance is not equivalent to the fact that these standards are no longer in force. The science system has its own standards of correct action, which are the measure of non-compliance. Does this mean, then, that we only have to insist that the standards are respected in order to restore the credibility of the players on the one hand and the system of science on the other?

I propose that this is exactly what should be done first, i.e. to refer to the central standards of these areas of action and to demand that they have to be respected. But that is not enough. In view of the increasing spread of bullshit, a more fundamental reflection on the status and role of science (and the media who communicate scientific results) in state and society is called for in a further step. One has to ask what the obligations of the respective agents are. Both the media and science have evolved over time from the actions and cooperation of individual agents. However, both areas of action now have a special legal status. This constitutional status in Germany (see footnote 1), which grants them a sphere largely free from state intervention, results from the special role that is attributed to both the media and science in a liberal democracy.[16] In a liberal democracy, the formation of the public will should take place with the participation of citizens. Both public debate and the provision of knowledge as the basis for action are regarded as necessary for the formation of opinion by each individual and, in particular, by policy-makers. In liberal democratic societies, it is considered that these goals are best served by guaranteeing the freedom of science and the press. In addition to the protection of freedom, however, the

---

16   The freedom of science from state intervention has constitutional status in the Federal Republic of Germany and some other states In other liberal democracies, e.g. in the USA, freedom of science is linked to the constitutional amendment on freedom of opinion. See Weingart 2014, sect. 3.

support for this freedom is also remarkable: not only is the freedom of science protected by the state, but science is also supported by the state. Universities and public research institutions offer scientists the opportunity to carry out research with due care and – this is at least an ideal – free from the worry of economic success.[17]

This granting of a sphere that is largely free of political intervention and its promotion, however, does not appear from nowhere, but on the basis of a functioning system of scientific norms that has evolved over centuries. It is the duty of scientists and philosophers of science to reflect on this system, to strengthen it and to make it comprehensible to a broader public not only in terms of its results, but also with regard to its system properties and here above all its process character. The Sars-CoV-2 pandemic in 2020 is an example that makes this necessity obvious. In this situation, scientific expertise provides the basis for political decisions, some of which are accompanied by considerable restrictions of fundamental rights. This makes it all the more important that scientists, when communicating scientific (interim) results, make it clear what epistemic status their utterance has and to what level of knowledge they refer. Otherwise, the changes in what is presented as – preliminary – findings and on which the restrictive measures are based cannot be made sufficiently clear to the public, i.e. to the persons suffering from them. It must be conveyed that the clinical picture of »Covid-19« is something that is not simply available, but that it is the result of a long and complicated and often controversial scientific process. From initial data, first conjectures about general features of the disease are generated, carrying out – correctly – weak affirmative acts. New data make it necessary to revise first conjectures, to make new assumptions, as well as to verify or falsify them. The road to a stable clinical picture is long. The example shows how important it is to provide an initial understanding of the standards-driven scientific process, i.e. knowledge generation.

In my opinion, reflecting on this system of scientific norms involves scientists of all disciplines, i.e. not only from the natural sciences, but also from the humanities, to continually check the objectives, objects and methods of their disciplines and set the standards for their statements and the claims that can be legitimately made. This also includes thinking about the relationship between the production or acquisition of knowledge, the internal and external communication of knowledge and recommendations for action. *What can be*

---

17   This idea is now being thwarted by the payment of performance bonuses etc. It remains an open question, which design of the remuneration system better promotes the production of substantial knowledge. Many questions about the topical direction of scientific work through third-party funding etc. could also be addressed.

*presented as knowledge with which degree of reliability and on which objectives
can recommendations be based?*

The demand for self-assurance is becoming more urgent in view of the
increasing number of newly emerging mixed disciplines, that integrate parts
of different scientific disciplines – data sciences, media studies, science and
technology studies, to name just a few.[18] Another *challenge* to the core norms
of gaining knowledge may come from a tendency that replaces

> »[...] the belief that science is an enterprise governed by universalistic norms
> that produces objective knowledge by abstraction from its bearers [...] by an
> individualization of the perception of reality and the production of knowledge«.
> (Weingart 2015, 13, presenting the view of Yacon Ezrahi)

In addition, the use of machine learning in pattern recognition, i.e. an even
greater significance of correlations than has already been brought about by
statistical methods, poses a further challenge for reflection and dissemination
of standards of producing truths.

A further element of self-assurance is the inclusion of the history of science.
The insight could consist in the fact that the system of science has, by and
large, produced results which provided a reliable basis for our further actions
even if an immediate usefulness was not always initially discernible (insights
in mathematics and logic at the beginning of the 20th century formed the
basis for later success in physics and current computer science could not
exist without this fundamental research). Obtaining evidence of such a link
between initially purposeless research and at least some fruitful results in the
long run reinforces the need for a »shelter« for science and provides pragmatic
arguments in favour of supporting basic research. – The practice of third-party
funding and the politically initiated founding of institutes with more or less
clear demands for problem solutions are omnipresent. Many scientists see
themselves dependent on this type of funding, also for the promotion of young
scientists. Nevertheless, the question can be raised: Would it not be appropriate to at least advocate strengthening non-project-related research funding?
Scientists could counter these attempts to steer research more convincingly, if
they adhere to their own standards.

The combination of scientific expertise and recommendations for action
deserves a special reflection on standards of correctness: Recommendations
for action require reference to objectives. What are the goals scientists assume

---

18   Weingart (2014, sect. 4) points to the problem with regard to the fact that in Germany it is
      left to the scientific community to determine what a scientific discipline is.

when they make recommendations?[19] Scientists *correctly* recommend a measure or an action if they *hypothetically* recommend: If you want to achieve this and that goal, then it is necessary to take measure A. Often this relationship between goal and action is not so simple. The pursuit of some goals with certain actions hinders other goals, in other words: there are conflicting goals that require weightings. If scientists perceive such conflicts of objectives on the basis of their expertise, they must also disclose them. The network of politically explicitly stated goals, possible measures of goal pursuit as well as the consequences for other, possibly implicitly pursued goals must be explained.

Politicians, on the other hand, have the task of dealing with these hypothetical recommendations. In other words, they have to decide which goals to weigh against possible consequences and they have to initiate the implementation of the corresponding measure supported by scientific expertise. The choice and weighting of objectives remain the responsibility of the decision-makers.

Scientists should take this type of division of labour into account if they want to remain trustworthy. Not to spread bullshit means: To present only that in the appropriate strength as an insight that is supported by the methods of one's own discipline and to give hypothetical recommendations that take into account the network of potentially pursued goals and possible consequences. To present this in sufficient transparency and to defend against further attacks seems to me to be one of the current challenges for scientists in public discourse.

References

Frankfurt, H.G. 2005. *On Bullshit*. Princeton: Princeton University Press.
Frankfurt, H.G. 2006. *On Truth*. New York: Knopf.
Hagendorff, T. 2020. »The Ethics of AI Ethics: An Evaluation of Guidelines.« Minds and Machines 30: 99–120. https://doi.org/10.1007/s11023-020-09517-8.

---

19   The discussion on the question of whether scientists may or should integrate non-scientific values and goals into their research was made famous by Max Weber. Weber limits the performance of science to the identification and provision of means-ends relationships, the indication of side effects in the pursuit of given goals and the resulting competition between values and goals. See Weber, [1917] 1988, 508. – It is the duty of researchers to explicitly point out the objectives they presuppose and the relativity of their recommendations. See Schurz 2013. – The logical expression of this connection is the introduction of the conditional: *If* one pursues the goal Z, *then* it is advisable to take the means M.

Künne, W. 1985. »Wahrheit.« In *Philosophie. Ein Grundkurs*, edited by E. Martens & H. Schnädelbach, 116–171. Hamburg 1985.

Michel, J.G. 2019. »How Are Species Discovered? Declarative Speech Acts in Biology.« *Grazer Philosophische Studien* 96 (3): 419–441.

Mill, J.S. [1861] 1977. *Considerations on Representative Government.* In *The Collected Works of John Stuart Mill, Volume 19 – Essays on Politics and Society, Part II*, edited by J.M. Robson, Introduction by A. Brady. Toronto: University of Toronto Press, London: Routledge and Kegan Paul. https://oll.libertyfund.org/title/robson-the-collected-works-of-john-stuart-mill-volume-xix-essays-on-politics-and-society-part-2#lf0223-19_head_008.

Schlick, M. [1918] 1974. *General Theory of Knowledge.* New York: Springer.

Schurz, G. 2013. »Wertneutralität und hypothetische Werturteile in den Wissenschaften.« In *Werte in den Wissenschaften: Neue Ansätze zum Werturteilsstreit*, edited by G. Schurz & M. Carrier, 305–334. Frankfurt: Suhrkamp.

Siegwart, G. 2007. »Alethic Acts and Alethiological Reflection.« In *Truth and Speech Acts: Studies in the Philosophy of Language*, edited by D. Greimann & G. Siegwart, 41–58. New York & London: Routledge.

Staffler, L. 2017. »Presseinterview als fahrlässige Tötung: Der italienische Strafprozess gegen die Expertenkommission zum Erdbeben von L'Aquila (2009).« *Zeitschrift für Internationale Strafrechtsdogmatik* 2: 125–138. http://www.zis-online.com/dat/artikel/2017_2_1090.pdf.

Johnston, B.C., Alonso-Coello, P., Bala, M.M. et al. 2018. »Methods for Trustworthy Nutritional Recommendations NutriRECS (Nutritional Recommendations and Accessible Evidence Summaries Composed of Systematic Reviews): A Protocol.« *BMC Medical Research Methodology* 18: 162. https://doi.org/10.1186/s12874-018-0621-8.

Weber, M. [1917] 1988. »Der Sinn der ›Wertfreiheit‹ der soziologischen und ökonomischen Wissenschaften.« In *Gesammelte Aufsätze zur Wissenschaftslehre.* 7th ed. Tübingen: Mohr Siebeck.

Weingart, P. 2014. »Die Stellung der Wissenschaft im demokratischen Staat. Freiheit der Wissenschaft und Recht auf Forschung im Verfassungsrecht.« *Zeitschrift für Theoretische Soziologie, Special Issue 2. Autonomie revisited: Beiträge zu einem umstrittenen Grundbegriff in Wissenschaft, Kunst und Politik*, edited by M. Franzen, A. Jung, D. Kaldewey, & J. Korte, 305–329.

Weingart, P. 2015. »Norms in Science.« In *International Encyclopedia of the Social & Behavioral Sciences*, vol. 17, 2nd. ed., edited by J.D. Wright, 11–14. Oxford: Elsevier.

Index